Aerodynamics of Wind Turbines

Emerging Topics

Aerodynamics of Wind Turbines

Emerging Topics

Editors

Ryoichi S. Amano
University of Wisconsin-Milwaukee, USA

&

Bengt Sundén
Lund University, Sweden

WIT*PRESS* **Southampton, Boston**

Editors:

Ryoichi S. Amano
University of Wisconsin-Milwaukee, USA

&

Bengt Sundén
Lund University, Sweden

Published by

WIT Press
Ashurst Lodge, Ashurst, Southampton, SO40 7AA, UK
Tel: 44 (0) 238 029 3223; Fax: 44 (0) 238 029 2853
E-Mail: witpress@witpress.com
http://www.witpress.com

For USA, Canada and Mexico

WIT Press
25 Bridge Street, Billerica, MA 01821, USA
Tel: 978 667 5841; Fax: 978 667 7582
E-Mail: infousa@witpress.com
http://www.witpress.com

British Library Cataloguing-in-Publication Data

A Catalogue record for this book is available
from the British Library

ISBN: 978-1-78466-004-8
eISBN: 978-1-78466-005-5

Library of Congress Catalog Card Number: 2014950374

Printed by Lightning Source, UK.

CONTENTS

CHAPTER 3

CHAPTER 4

PREFACE

This book presents some emerging topics on aerodynamics of wind turbines. The topic ranges from fundamental to application of dynamics of wind turbines, flow and aerodynamics of wind turbine blades, and blade designs.

The first chapter presents the introduction of wind energy and energy resources, the world trend of wind energy usage and some basic concepts of the wind turbines. It is presented that technological development of recent years, bringing more efficient and more reliable wind turbines, is making wind power more cost-effective. In general, the specific energy costs per annual kWh decrease with the size of the turbine notwithstanding existing supply difficulties.

The second chapter covers some basic theories associated with wind turbines. The readers can review basic concept and the theories used to understand the wind turbine aerodynamics and wind resources. The theory for performance of a wind turbine is introduced that includes the power, torque and thrust. This chapter's main emphasis is the basic theory on determining the amount of energy captured by the rotor, the rotor thrust, power and non-dimensional-characteristic performance curves from which the actual performance can be determined. The Blade Element Momentum (BEM) theory, full computer simulation using Reynolds-averaged Navier–Stokes (RANS) method, Prandtl's analysis and advanced turbulence models are discussed in some details.

The third chapter describes an automatic control and condition monitoring of wind turbines, which are usually treated as two distinct technical areas because control has a short-term focus whereas condition monitoring has a long-term focus. First, the vibrational characteristics of the rotor system are described in the context of modal analysis where each mode of vibration occurs at a particular frequency with a particular modal deflection shape. Secondly, the effects of changes in the wind state on the sensitivity of condition monitoring data to damages in the turbine rotor are discussed. It is shown using modal filters that certain frequencies in the measured response of the wind turbine are sensitive to a change in the condition of the rotor in the form of ice accretion or blade root damage for use in detecting that type of damage. By analyzing the changes in the modal response amplitudes, the type of damage that is present can be classified.

The fourth chapter focuses mainly on the future of wind turbine design and the wind tunnel testing to produce high quality, experimental data used for design. These experimental data, both aerofoil and overall system performance, can be used to validate and improve the design of the wind turbine blades and systems. Currently, little experimental testing of wind turbines is accomplished, with much of the aerodynamic design being accomplished using computational fluid dynamic

(CFD) tools. CFD prediction is improving and will be an outstanding tool for wind turbine blade design; however, these codes are not robust enough for predicting performance under low Reynolds numbers. To support the effort of developing CFD modeling, the performance of the full-scale wind turbine systems is monitored and the results are presented for the wind turbine design. This chapter emphasizes the necessity of experimentally testing wind turbine blades to determine aerofoil lift and drag data over typical Reynolds numbers of operation and also the testing of wind turbine systems (blades and generator) to determine the overall wind turbine performance. This type of testing should be accomplished prior to building the full-scale machine. Also, for small-scale wind turbines, the importance of quality experimental aerofoil data taken at the appropriate Reynolds numbers is necessary for accurate design and prediction of power production. The data presented are for wind tunnel testing in air; however, there is a growing interest in testing wind turbines in water tunnels, although not many facilities exist for this application. The discussions presented should be applicable to either air or water testing.

The fifth chapter presents how the blade shape affects the overall efficiency and performance of the unit, which are of great interest to manufactures and their customers alike. Since blades of commercial wind turbines can reach upwards of 20 m in length, the physical test of various designs can become expensive and time consuming. Although there are some general analytical formulas and procedures for determining the blade shape, these tend to fall short of producing optimal designs given the several significant in accurate assumptions their derivation is founded upon. It is for this reason researches have turned to computer simulation to develop and optimize wind turbine blade shape. CFD is a powerful tool that has been leveraged in many industries including the wind power industry. When applied and executed correctly, simulation results very closely approximate actual blade performance and thus can be used to evaluate and identify optimal designs that will perform up to expectations in the real world.

The sixth chapter discusses how to find out the influence of the ice accretion on the wind turbine blade performance, to reveal how different icing variables affect ice accretion and to estimate the heat amount required for de-icing the blades at different icing events. Also different computational model simulations of ice accretion and the different de-icing technologies existing today are described. The study emphasizes the conditions that are most severe for ice accretion. However, the chapter is theoretical, so experiments are needed to validate and complement the conclusions reached in this chapter. It is pointed out that it is important to find the conditions for which ice accretion is formed and the heat amount required to avoid it, to design suitable blades for regions where ice accretion is formed regularly and develop suitable de-icing systems.

The seventh chapter states the methods for a blade breakage self-repair system with an innovative self-healing approach. To prevent blade breakage due to excessive wind speed, the methodology of self-healing technology can possibly be implemented with a self-healing wind turbine blade – a new and innovative concept for wind turbine blade design. The composite blades are manufactured using a laborious, hand lay-up technique with glass fibre reinforcements. The use of fibre-reinforced composite materials has grown rapidly and such composites are often used in aerospace and other applications; however, concerns remain about the structural integrity of composite materials following impact loading, as they are susceptible to cracks or delaminations that form deep within the structure. With the success of the self-healing technology for wind turbine blades, any cracked parts in wind turbine blades can be healed during operation without any system shutting down. Therefore, technology sheds light on the existing wind turbine failure prevention methods and provides new concepts and approaches for new materials of turbine blades suitable for the 21st century wind energy era.

All of the chapters follow a unified outline and presentation to aid accessibility and the book provides invaluable information to both graduate researchers and industrial research engineers/scientists.

We are grateful to the authors and reviewers for their excellent contributions. We also thank ceaseless help that were provided by the staff members of WIT Press, in particular Mrs. Elizabeth Cherry, for their strong encouragement in the production of this book. Finally, our appreciation goes to Dr. Carlos Brebbia who gave us strong support and encouragement to complete this project.

Ryoichi S. Amano and Bengt Sundén

ABOUT THE EDITORS

Professor Ryoichi S. Amano is an internationally recognized scientist in thermal engineering, experimental fluid dynamics, turbulence research, and energy systems. For a number of years he has been engaged in research related to gas turbine performance, rotating machinery, power engineering, transport phenomena, heat/mass transfer, two-phase flow, and manufacturing material processes. Dr Amano has also contributed to the development of turbulence theories, jets, combustion, heat transfer, propulsion, aerodynamics, and applications to gas turbine and aerospace-related projects. In addition, he has engaged in gas turbine flow analyses of NASA space shuttle main engines and solid rocket motor research of US Air Force Research Labs, industrial gas turbines, steam turbines, and cooling technology using impinging gas jets.

Dr Amano has more than 500 publications, including books, refereed journal papers, invited review chapters, and conference proceedings. He has conducted numerous experimental and analytical research projects with extramural funding from US governmental agencies and many industries. Dr Amano is a technical committee member, editorial board member and international advisory committee member for ASME, AIAA, WIT Conferences, ISETS conferences (Japan) and an executive member of ASME International Gas Turbine Institute, AIAA Terrestrial Energy System, and ASME Energy Systems Committee. In addition, Dr Amano has received three Best Paper awards from ASME, the Sustained Service Award from AIAA, the AIAA Energy Systems Award, the ASME George Westinghouse Gold Medal, and a UWM Excellence of Research Award. He is a Fellow of ASME and Associate Fellow of AIAA.

Professor Bengt Sundén received his MSc in 1973, PhD in thermodynamics and fluid mechanics in 1979, and Docent in applied thermodynamics and fluid mechanics in 1980, all from Chalmers University of Technology, Goteborg, Sweden. He became Professor of Heat Transfer in 1992 at Lund University. Since 1995 he serves as the head of the Department of Energy Sciences, Lund University, Sweden, and is also a guest professor at Northwestern Polytechnical University, Xi'an, China, and an honorary professor at Xi'an Jiatong University, Xi'an China.

His research topics include compact heat exchangers, enhancement of heat transfer, gas turbine heat transfer, combustion-related heat transfer including thermal radiation, CFD-methods for laminar and turbulent fluid flow and heat transfer, liquid crystal thermography, condensation and evaporation, nanofluids, transport phenomena in fuel cells, computational modeling and analysis of multiphysics and multiscale phenomena for fuel cells (SOFC, PEMFC).

Professor Sundén was the founding and first editor-in-chief of IJHEX (International Journal of Heat Exchangers), and was Associate Editor ASME Journal of Heat Transfer. In addition, he is currently an active editor for three journals. He is also editor-in-chief for a book series, Developments in Heat Transfer (also published by WIT Press), has published over 600 papers in journals, books and conference proceedings, and has supervised 170 MSc theses, 44 Licentiate of Engineering theses and 40 PhD theses. According to ISI knowledge Web of Science, the overall number of citations is 35 per year and the total number of citations is more than 2300. The h-index is 25.

He is a fellow of the ASME and a 2011 recipient of the ASME Heat Transfer Memorial Award, and a 2013 recipient of the ASME Heat Transfer Division 75th Anniversary Medal. He is a fellow of the Wessex Institute of Technology and a holder of the WIT Eminent Scientist Medal.

CHAPTER 1

Introduction to Wind Power

R.S. Amano
University of Wisconsin-Milwaukee, Madison, WI, USA.

Abstract

Wind power is a well-proven and cost-effective technology and expected to be the main way in which industry responds to the Government's targets – so becoming an important source of electricity in years to come. In this book, we have focused on basic and some advanced topics needed to understand the current situation in the win power engineering. This chapter provides introductory materials for the wind energy.
Keywords: Introduction to renewable energy, wind energy in the world, fundamentals in wind turbines.

1 Introduction

As the clean energy industry emerges from initial stage caused by the global economic downturn, it is entering a new stage of rapid change of business. The worldwide demand for energy is expected to double by the year 2030 and triple by 2050, when fossil fuels will account for no more than two-thirds of all energy consumed, compared with 79% of the energy consumed today. Traditional fossil sources such as oil, gas and coal are not renewable and cause pollution by releasing huge quantities of carbon dioxide and other pollutants into the atmosphere, thereby damaging the environment in many ways, from acid rain to climate change. To help combat these problems, many states in the United States are seeking ways to use renewable energy sources, such as wind, solar and biomass. Along with its environmental and cost benefits, renewable energy is a rapidly growing industry with vast potential for economic growth and job creation. In fact, the U.S. Secretary of Agriculture has identified wind, solar and biomass as key factors for advancing the U.S. economy.

Wind energy has recently become the world's fastest growing source of renewable energy. The U.S. Department of Energy (DOE) expects that wind energy will contribute to 20% of the U.S. electricity supply by 2030. As a result, there has been a revived interest in wind turbines because they are emissions-free and wind is renewable and cost-free; however, the amount of electricity generated and obtained by wind energy conversion systems is still unsteady, relatively expensive and difficult to integrate into traditional electricity systems because of the variation in wind source and unresolved energy storage issues. On a large scale, spatial variability describes the fact that there are many different climatic regions in the world, some much windier than others. These regions are largely dictated by the latitude, which affects the amount of insolation. Within any one climatic region, there is a great deal of variation on a smaller scale, largely dictated by physical geography – the proportion of land and sea, the size of land masses and the presence of mountains or plains, for example. The resource map of wind energy in the United States [1] indicates that the vast majority of available wind is very unsteady; strong wind zones are concentrated in certain regions and not uniformly distributed throughout the nation, making wind power collection more difficult. Conversely, the easy-to-collect wind energy is primarily confined to remote locations, making electricity distribution difficult.

In the USA, while wind makes up only 2% of total electricity supply, it is one of the largest sources of new power generation in the country, second only to natural gas generation in terms of new capacity built each year since 2005. During the last few years, wind power in the United States has been increasing rapidly as shown in Fig. 1 [2]. Over 1,000 wind turbines >2 megawatts (MW) are already in commercial operation in the United States, and the year-end order for 338 GE 2.5-MW wind turbines for the Shepherd's Flat wind project in Oregon is the harbinger of a shift in orders towards such larger turbines. This forecast assumes that inventory will have been exhausted and that there will be a growing market for wind turbine orders in 2011 and beyond, spurred by a national Renewable Electricity Standard (RES). The trend towards larger turbines is driven by economics: taller turbines with larger swept areas produce more power at a lower cost per kilowatt-hour. The rapid quest for more electrical power, moving from 2 MW turbines in 2009 to 6 MW turbines being installed in early 2012 has led to the tower head mass being increased from 140 to 360 tonnes, with enhanced variations in structural loading and fatigue. Stabilisation at the current power level will enhance opportunities for consistent design and manufacture of the structures within a farm.

Total wind capacity in the USA reached 60,000 MW by the end of 2012, with commercial-scale wind turbines operating in 38 states. Wind power accounted for 35% of the country's new power-production capacity from 2007 to 2011, second only to natural gas. According to the American Wind Energy Association (AWEA), Texas leads the country as the state with the most installed wind power with 10,400 MW. Iowa remains a leader in wind generation with 3,675 MW installed; while California and Minnesota continue to harvest significant amounts of wind with 4,322 MW and 2,733 MW, respectively [3].

Although this is a significant growth for wind energy, it still only accounts for a small percentage of the U.S. electricity supply. The U.S. Department of Energy recently released a report that laid out a plan to reach 20% wind energy power by

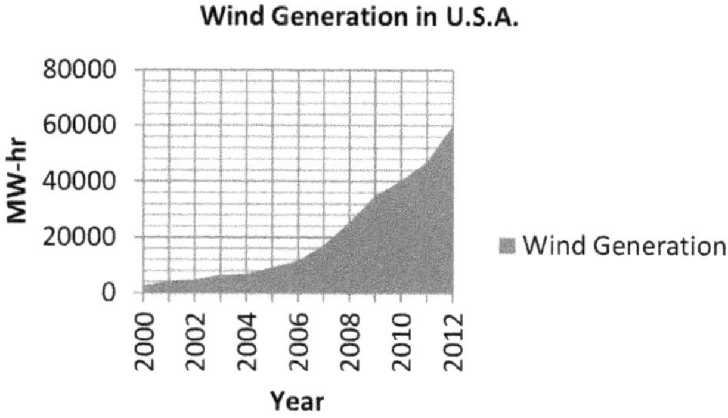

Figure 1: U.S. wind power generation (MW-hr).

2030 to fuel the U.S. electricity grid. This would provide a major increase in jobs, benefits to rural landowners and lead the Country to increased energy independence. Factors pushing for growth in U.S. wind energy include the high cost of fossil fuels and concern over national energy security. As a result, policy makers are actively considering a wide range of legislation that would support and enhance wind energy growth.

Progressive public policy has usually been a key ingredient both for encouraging wind energy expansion and helping to determine what forms that growth will take. Future growth will likely come from commercial-scale wind farms, which are typically vast arrays of turbines owned and operated by large corporations. Yet experience in Minnesota has shown that, with an encouraging policy environment, small clusters of turbines or even single turbines can make significant contributions, operated by local landowners, small businesses and community wind projects.

Worldwide there are now over two hundred thousand wind turbines operating, with a total nameplate capacity of 282,482 MW as of end 2012 [4] The European Union alone passed some 100,000 MW nameplate capacity in September 2012 [5], while the United States surpassed 50,000 MW in August 2012 and China passed 50,000 MW the same month [6]. World wind generation capacity more than quadrupled between 2000 and 2006, doubling about every 3 years. The United States pioneered wind farms and led the world in installed capacity in the 1980s and into the 1990s. In 1997, German installed capacity surpassed the United States and led until once again overtaken by the United States in 2008. China has been rapidly expanding its wind installations in the late 2000s and passed the United States in 2010 to become the world leader.

At the end of 2012, worldwide nameplate capacity of wind-powered generators was 282 gigawatts (GW), growing by 44 GW over the preceding year [7]. According to the World Wind Energy Association, an industry organisation, in 2010 wind power generated 430 TWh or about 2.5% of worldwide electricity usage [8] up from 1.5% in 2008 and 0.1% in 1997 [9]. Between 2005 and 2010, the average annual growth in new installations was 27.6% [10]. Wind power market penetration is expected to reach 3.35% by 2013 and 8% by 2018 [10]. Several countries

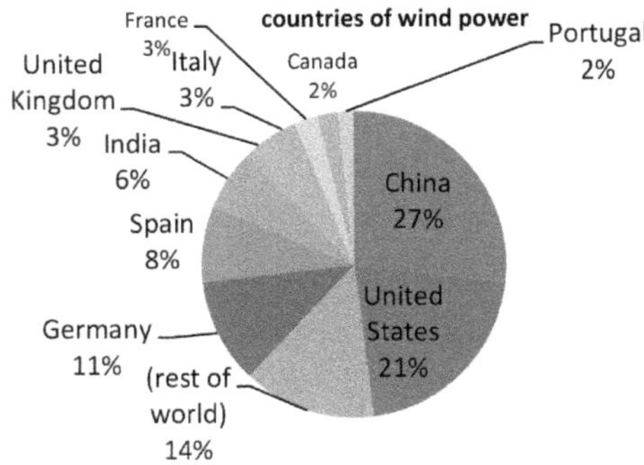

Figure 2: Top 10 countries by name plate wind power capacity (2012 year-end).

have already achieved relatively high levels of penetration, such as 28% of stationary (grid) electricity production in Denmark (2011) [11] 19% in Portugal (2011), 16% in Spain (2011), 14% in Ireland (2010) and 8% in Germany (2011). As of 2011, 83 countries around the world were using wind power on a commercial basis [12]. Figure 2 shows the top 10 countries by nameplate windpower capacity by the end of 2012.

2 Why do we need wind energy

The world is getting hotter, in fact by 1°C on land over the last 100 years, and the overwhelming consensus of scientific opinion is that human activities, particularly the emission of greenhouse gasses, are the cause [13]. The energy sector is by far the biggest source of these emissions, and if we are to tackle climate change it is clear that we need to move away from burning limited fossil fuel reserves to more sustainable and renewable sources of energy. As well as being good for the planet, this is also good for economy as it reduces the dependency on oil and gas imports – improving the balance of payments while also increasing energy supply security. We need to see more renewable energy and the Government has set a binding target for 20% of the USA's electricity to come from renewable sources by 2030 [14].

Wind power can play a major role in meeting USA's increasing demand for electricity, according to a ground-breaking technical report, 20% Wind Energy by 2030: Increasing Wind Energy's Contribution to U.S. Electricity Supply, prepared by the U.S. Department of Energy with contributions from the National Renewable Energy Laboratory, the AWEA, Black & Veatch and others from the energy sector. To implement the 20% Wind Scenario, new wind power installations would increase to more than 16,000 MW per year by 2018, and continue at that rate through 2030, as shown in Fig. 3. Wind plant costs and performance are projected to improve modestly over the next two decades, but no technological breakthroughs are needed. In the 20% wind scenario, 46 states would experience significant wind power development.

The report finds that during the decade preceding 2030, the U.S. wind industry could:

(a) support roughly 500,000 jobs in the United States, with an annual average of more than 150,000 workers directly employed by the wind industry;
(b) support more than 100,000 jobs in associated industries (e.g. accountants, lawyers, steel workers and electrical manufacturing);
(c) support more than 200,000 jobs through economic expansion based on local spending;
(d) increase annual property tax revenues to more than $1.5 billion by 2030 and
(e) increase annual payments to rural landowners to more than $600 million in 2030.

The report explores one scenario for reaching 20% wind electricity by 2030 and contrasts it to a scenario in which no new U.S. wind power capacity is installed. It examines costs, major impacts and challenges associated with the 20% Wind Scenario. It investigates requirements and outcomes in the areas of technology, manufacturing, transmission and integration, markets, environment and siting. The report finds that the Nation possesses affordable wind energy resources far in excess of those needed to enable a 20% scenario [15].

Wind power is a well-proven and cost-effective technology and expected to be the main way in which industry responds to the Government's targets – so becoming an important source of electricity in years to come.

Wind farms are created when multiple wind turbines are placed in the same location for the purpose of generating large amounts of electric power. Due to rising energy prices and the resultant search for alternatives, there are now thousands of wind farms in many countries around the world. There is still a lot of controversy surrounding the pros and cons of wind power and its local impact. The articles listed on this page explore news and information about wind farms. Now major electric companies are going green and proudly proclaiming it too from rooftops.

3 Current status on wind energy technology

Over the two decades, average wind turbine ratings have grown almost linearly seven times as large as 1980s (see Fig. 3). Each group of wind turbine designers has predicted that their machines are as large as they will ever be. However, with each new generation of wind turbines, the size has increased along the linear curve and has achieved reductions in life-cycle cost of energy. The long-term drive to develop larger turbines stems from a desire to take advantage of wind shear by placing rotors in the higher, more energetic winds at a greater elevation above ground (wind speed increases with height above the ground). For this reason, the capacity factor of wind turbines has increased. However, there are constraints to this continued growth to larger sizes as in general it costs more to build a larger turbine. The primary argument for a size limit for wind turbines is based on the 'square-cube law'. Roughly stated, it says that 'as a wind turbine rotor increases in size, its energy

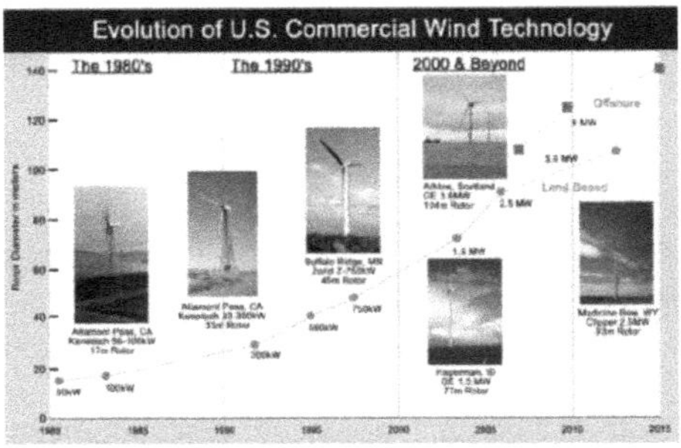

Figure 3: The turbine blade size growth over 25 years [16].

output increases as the rotor-swept area (the diameter squared), while the volume of the material, and therefore its mass and cost, increases as the cube of the diameter'. In other words, at some size the cost for a larger turbine will grow faster than the resulting energy output revenue, making scaling a losing economic game. Engineers have successfully skirted this law by changing the design rules with increasing size and removing material or by using material more efficiently to trim weight and cost. Studies have shown that in recent years, blade mass has been scaling at roughly an exponent of 2.3 instead of the expected 3, as shown by the WindPACT blade scaling study. The WindPACT study shows how successive generations of blade design have moved off the cubic weight growth curve to keep weight down as illustrated in Fig. 3. If advanced research and development were to provide even better design methods, as well as new materials and manufacturing methods that allowed the entire turbine to scale as the diameter squared, then it would be possible to continue to innovate around this limit to size [16].

4 VAWT and HAWT

Cross-wind-axis wind turbines most commonly oriented vertically and termed Vertical Axis Wind Turbines (VAWTs) have emerged as a potential unit to our need, after being shelved by wind turbine companies in the late 1980s as a result of the greater success of Horizontal Axis Wind Turbines (HAWTs). VAWTs present several advantages over HAWTs, however, which are especially pertinent in the built environment. There are, however, also significant challenges that have prevented their widespread adoption. These challenges must be overcome if VAWTs are to significantly contribute to meeting the 2030 wind power goal of the Department of Energy.

VAWT's have the equipment that needs maintenance nearer the ground. VAWT's are smaller (because if they were made bigger they would fly apart) and are used only in small applications – homes, office buildings, etc. where the reason is to make a statement. They make less than 50 kW. There are no commercial-sized VAWT's, because they do not make money, but if they existed they would spin at speeds from 1 mph to 100s of miles per hour. VAWTs would, of course, fly apart as soon as the wind went over about 20 mph, which is why they are not in commercial operation, as centripetal forces would break the blades. There are lots of types of VAWT's, because they are really all 'experimental' in nature, although some are in mass production. VAWTs cost less in total dollars, but significantly more in dollars per kilowatt generated (because they do not make much power for their cost).

4.1 VAWT drawbacks

Although several large test facilities have been built, these are for the most part no longer in operation. In addition to the principal advantages of wind turbines with a vertical axis – for instance, the low levels of noise and the independence from the wind direction – there are unfortunately several disadvantages. One of these is that they must be installed close to the ground. Since the wind naturally blows more strongly and evenly at greater heights, an installation that is not on a mast loses a great deal of efficiency. If with this type of installation, the generator is housed in a machine room on the ground, then the maintenance is of course simpler and cheaper. In spite of this, it is doubtful that the lower yield because of the weaker winds close to the ground would be balanced out by the money saved in maintenance costs. It still remains to be seen whether plans to use existing tall structures to mount planned megawatt-level installations with vertical spindles can be realized. It will probably not be easy to find buildings or structures that would be able to handle the static and dynamic loads from a large wind power installation with a vertical axis. And it goes without saying that these structures should be in regions where the wind velocities are of interest. Another point against the current conceptions of larger VAWTs is the greater material expenditure per square meter of surface covered in comparison to installations with a horizontal spindle. This is a significant additional cost factor that can hardly be compensated for by the theoretically better possible exploitation of strong winds or gusts.

The main disadvantage of VAWTs is their low efficiency relative to HAWTs. This is a result of the variable torque produced by dynamic stall experience by each blade as it passes around the azimuth. The blades of a HAWT, on the other hand, produce constant torque around the azimuth. A further disadvantage is the inability to VAWTs to self-start. By improving the torque and therefore power produced by VAWTs, and enabling self-start, the two main roadblocks to their adoption would be removed. This would enable the widespread adoption of VAWTs for low-wattage power generation in the urban and residential wind power markets. This will increase renewable power generation in the locations where most power is consumed, reducing the losses and costs associated with transmission. Furthermore, it will reduce the reliability on fossil fuel based power generation and relieve consumers from the price variations that result from this reliance.

4.2 VAWT advantage

VAWTs have several advantages over HAWTs in the urban environment. Because they operate with a lower tip speed than HAWTs, they are less noisy. They also

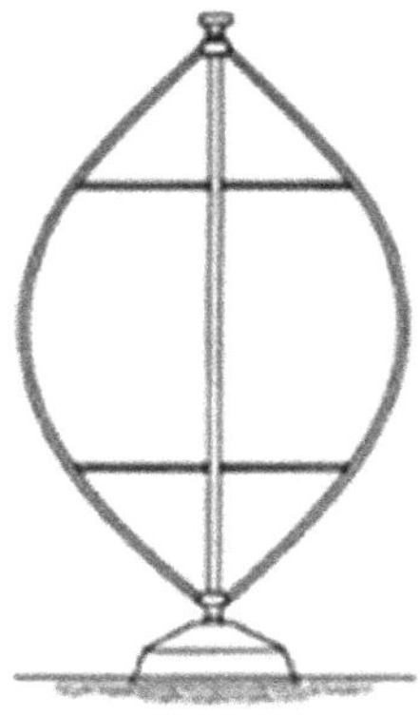

Figure 4: H-Darrieus VAWT.

have a better aesthetic due to their three-dimensional shape, making them popular with architects. HAWTs are sensitive to yaw and skew, experiencing decreases in torque and power due to the aerodynamic asymmetry on the rotor disc under such flow conditions. Due to the complexity of airflow in the urban environment, the wind direction is generally not perpendicular to the vertical, making this issue an important problem in this environment. VAWTs, however, are less sensitive to both yaw and skew [17]. A further advantage of the VAWT is the simplicity of the mechanical design and maintainability of the turbine system. Being based on a vertical shaft, VAWTs can be designed with all heavy components located at ground level, with the exception of a single bearing. This advantage is of particular interest to the residential power generation market, in which ease of repair is critical. Of the various configurations of VAWT that exist, the simplest is the H-Darrieus VAWT, illustrated in Fig. 4. One of its advantages is its ease of manufacture, since the blades can be extruded. This makes the H-Darrieus VAWT, in particular, less expensive to build. Even the curved and helical blades of the more common 'egg-beater' and helical VAWT designs are economic to manufacture than the tapered blades of HAWTs. By reducing the manufacturing cost of wind turbines for the home market, the installation cost is reduced, as is the return time on investment. This time is one of the main barriers to most renewable energy system in the home power market. On the other hand, the situation with smaller installations with a nominal output up to approximately 10 kW can be considered to be substantially different. At this level of output, there are very many applications, which until now could only be insufficiently covered with horizontal systems. In particular, horizontal installations come up against their limits when located in high mountain areas, in regions with extremely strong or gusty winds, or in urban areas. But also in regions with relatively constant winds, that is, where the conditions are ideal for systems with a horizontal axis, a VAWT can have its advantages, at minimum if the neighbours complain about the annoyance of the noise. There have already been reports of enraged neighbours who have settled the acoustical problems with firearms.

5 Conclusions

The use of wind energy was an old practice since thousands of years ago. However, this technology has revived due to the shortage of fuels and the environmental problems generated by the traditional energy resources. The last decade has seen a sharp increase in wind turbine generated electricity globally and is now well accepted with a large industry manufacturing and installing thousands of MWs of new capacity each year. Although there are exciting new developments, particularly in very large wind turbines, and many challenges remain, there is a considerable body of established knowledge concerning the science and technology of wind turbines. This book is intended to present some of this knowledge and to present it in a form suitable for use by students and by those involved in the design, manufacture or operation of wind turbines.

References

[1] Utility-Scale Land-Based 80-Meter Wind Maps. *Wind Power America*. U.S. Department of Energy—Energy Efficiency & Renewable Energy, 2013.
[2] American Wind Energy Association (AWEA). AWEA 3rd quarter 2012 Public Market Report, 2012.
[3] Wind Power in the United States. http://en.wikipedia.org/wiki/Wind_power_in_the_United_States
[4] GWEC Global Wind Statistics 2012. *Global Wind Energy Commission*. Retrieved 18 February 2013. 2012.
[5] EU wind power capacity reaches 100GW. *UPI*. October 2012. Retrieved 31 October 2012.
[6] China's on-grid wind power capacity grows. *China Daily*. 16 August 2012.
[7] Wind Power by Country. http://en.wikipedia.org/wiki/Wind_power_by_country. 2013
[8] World Wind Energy Report 2010. *World Wind Energy Association Report*. 2011.
[9] Wind Power Increase in 2008 Exceeds 10-year Average Growth Rate. *Worldwatch.org*. August 2010.
[10] BTM Forecasts 340-GW of Wind Energy by 2013. *Renewableenergyworld.com*. August 2010.
[11] Månedlig elforsyningsstatistik. *Danish Energy Agency*. January 2012. Vols. summary tab B58-B72.
[12] Wind Power. *Gree and Independent Future*. http://windivaenergy.com/index.php/wind-energy, 2008.
[13] Coriolis Energy. http://coriolis-energy.com/wind_energy/wind_index.html. 2013.
[14] 20 percent wind energy by 2030. http://www.20percentwind.org/20p.aspx?page=Overview. 2013.
[15] Energy efficiency and renewable energy. http://www.nrel.gov/docs/fy08osti/41869.pdf. 2013.

[16] Thresher, R., Robinson, M. & Veers, P., *Wind Energy Technology: Current Status and R&D Future*. National Renewable Energy Laboratory. 2008, NREL/CP-500-43374.
[17] Ferreira, C., Simão, J., *et al.,* Simulating dynamic stall in a 2D V modeling strategy, verification and validation with particle image velocimetry data.

CHAPTER 2

Basic Theory for Wind Turbine Blade Aerodynamics

P.M. Mohan Das & R.S. Amano
University of Wisconsin-Milwaukee, Madison, WI, USA.

Abstract

The performance of a wind turbine includes the power, torque and thrust. This chapter describes the basic theory on determining the amount of energy captured by the rotor. The rotor thrust has great influence on the structural design of the tower. It is usually convenient to express the performance by means of non-dimensional, characteristic performance curves from which the actual performance can be determined regardless of how the turbine is operated, for example, at constant rotational speed or some regime of variable rotor speed. The blade element momentum theory, full computer simulation using Reynolds-averaged Navier–Stokes method, Prandtl's analysis and advanced turbulence models are discussed.

Keywords: BEM, CFD, advanced flow simulations.

1 Introduction

A wind turbine is a device, which extracts kinetic energy from the wind and converts it to torque at the shaft. Traditional wind turbines can be broadly classified into lift machines and drag machines. Drag machines typically make use of the drag force produced by wind to generate power, whereas lift machines utilise lift force to generate torque and hence power. Considering that modern aerofoils at appropriate angle of attacks have a lift to drag ratio of up to 200, lift machines usually have a much higher coefficient of power compared with drag machines.

Typical methods available for aerodynamic modelling of wind turbines include those based on blade element momentum (BEM) theory, Navier–Stokes solvers or a hybrid methodology.

2 General methods

A broad understanding of the available energy and modes of operation can be obtained based on momentum theory based analysis of a wind turbine system.

2.1 Momentum theory

Momentum theory refers to the control volume analysis of the forces at the blade based on the conservation of linear and angular momentum, whereas blade element theory refers to the analysis of forces at a section of the blade, as a function of blade geometry.

As mechanical energy can only be extracted from a change in kinetic energy and hence velocity of the wind stream, the wind velocity behind the wind turbine has to be less than the wind velocity in front of the wind turbine. This would imply an increase in area since the density change is insignificant. Utilising an actuator disc and stream-tube assumption as shown in Fig. 1, we can estimate the power conversion from a wind turbine.

The mechanical energy that is extracted can be found by subtracting the available kinetic energy after the actuator disc from the initial free stream kinetic energy

$$P = \frac{1}{2}\rho({v^3}_1 A_1 - {v^3}_2 A_2). \tag{1}$$

By continuity equation,

$$\rho v_1 A_1 = \rho v_2 A_2. \tag{2}$$

Thus, eqn (2) yields as

$$P = \frac{1}{2}\rho v_1 A_1\left({v^2}_1 - {v^2}_2\right). \tag{3}$$

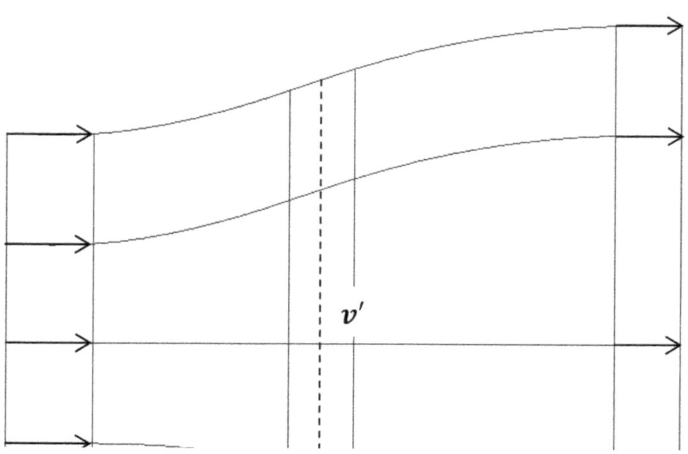

Figure 1: Actuator disc and stream tube [1].

$$P = \frac{1}{2}\dot{m}(v^2_1 - v^2_2) \quad (4)$$

As per eqn (5), for power extracted to be maximum, v_2 has to be zero. But this is physically impossible since it would imply no flow across the turbine.

Using conservation of momentum, the force exerted by the air on the disc can be expressed as

$$\bar{F} = \dot{m}\left(v_1 - v_2\right). \quad (5)$$

Assuming that the velocity of air in the disc plane to be v', the power required for moving the air at this velocity from the disc plane is

$$P = \bar{F}v' = \dot{m}(v_1 - v_2)v'. \quad (6)$$

By combining eqns (5) and (7),

$$v' = \frac{1}{2}(v_1 - v_2)\cdot \quad (7)$$

Hence,

$$\dot{m} = \frac{1}{2}\rho A\left(v_1 + v_2\right). \quad (8)$$

And mechanical power output forms the disc

$$P = \frac{1}{4}\rho A\left(v_1 + v_2\right)\left(v^2_1 - v^2_2\right). \quad (9)$$

The coefficient of power C_p is defined as the ratio of extracted power to the available power.

$$C_p = \frac{P}{P_0} = \frac{(1/4)\rho A(v_1 + v_2)(v^2_1 - v^2_2)}{(1/2)\rho v_1^3 A}. \quad (10)$$

By rearranging the terms,

$$c_p = \frac{P}{P_0} = \frac{1}{2}\left[1 - \left(\frac{v_2}{v_1}\right)^2\right]\left[1 + \frac{v_2}{v_1}\right]. \quad (11)$$

The induction factor 'a' is defined as

$$a = \frac{v_2}{v_1}. \quad (12)$$

The torque coefficient is estimated as

$$C_T = \frac{\text{Power}}{(1/2)\rho v_1^2 A} = 4a(1-a). \tag{13}$$

2.2 Betz limit

For maximum power extraction, $dc_p / d(v_2 / v_1)-$ has to be zero, which implies for maximum power output

$$\frac{v_2}{v_1} = \frac{1}{3}. \tag{14}$$

and

$$c_p = \frac{16}{27} = 0.593. \tag{15}$$

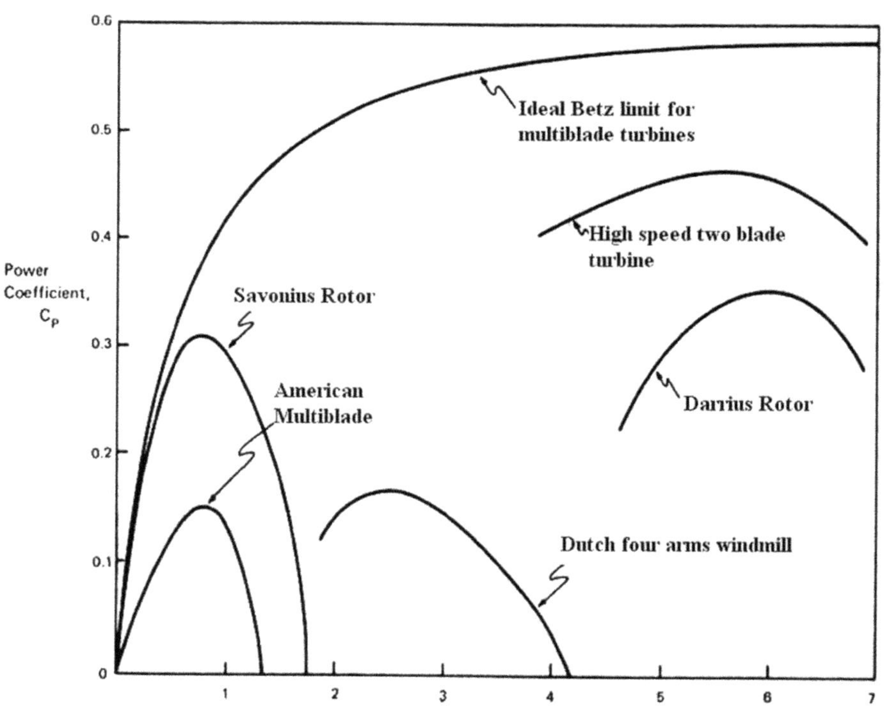

Figure 2: Rotor power coefficient variation with tip speed ratio for different types of wind turbines [1].

This maximum value for coefficient of power , which is the ability of wind turbine to extract energy from free stream wind was first predicted in 1915 by Lanchester [2]. In 1920, both Betz and Joukowsky derived this maximum efficiency independently of each other and unaware of Lanchester's work. See Fig. 2 for rotor power coefficients for other wind turbines. This limit has been popularly known after Betz and it is, therefore, frequently called as 'Betz factor' or 'Betz limit'. Betz limit indicates the physically based, ideal limit value for the extraction of mechanical power from free stream airflow without considering the design of the energy converter. In practical cases, a rotating energy converter will additionally impart a rotating motion, a spin, to the rotor wake. To maintain the angular momentum, the spin in the wake must be opposite to the torque of the rotor. The energy used for this spin reduces the useful portion of the total energy content of the air stream. This further reduces the coefficient of power from Betz's limit and power coefficient now becomes dependent on the ratio of the energy components from the rotating motion and the translatorial motion of the airstream. This available power is further dependent on the ratio of the rotational velocity of the blade to the free stream wind velocity. This ratio is called the tip speed ratio λ

$$\lambda = \frac{\Omega R}{V_1}. \quad (6)$$

2.3 States of operation of wind turbine

Typical wind turbine operation happens at or close to design or optimum conditions for power extractions. However, other states of operation can also occur. The

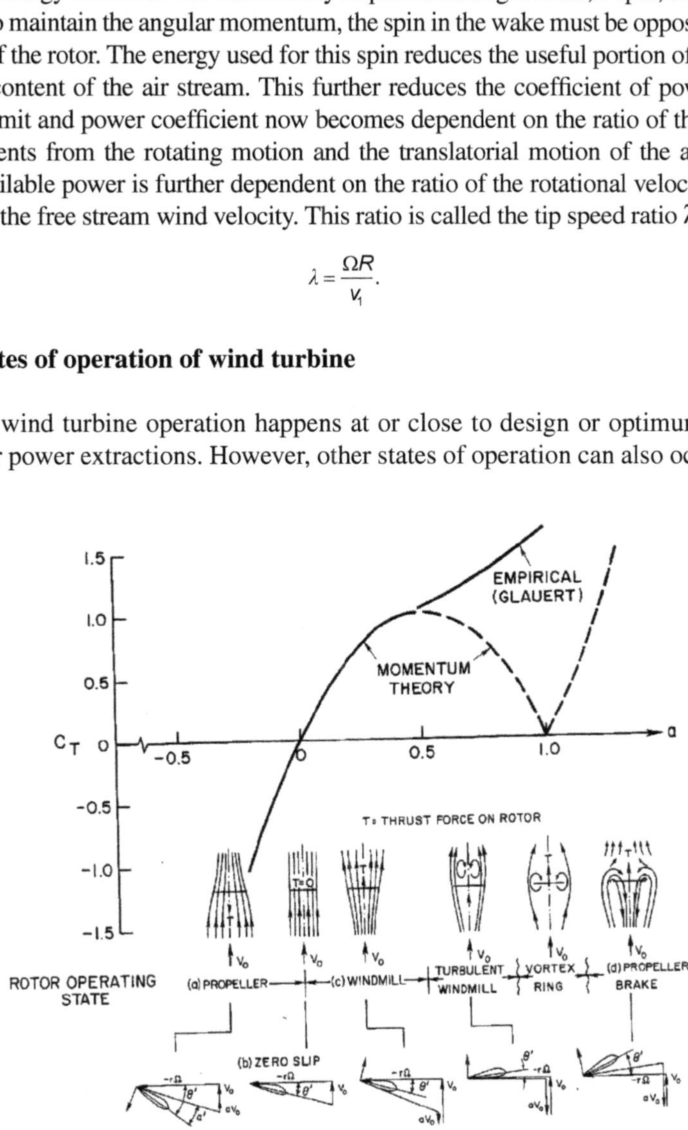

Figure 3: Rotor states [5].

theory of the states of operation of a rotor was first proposed by Glauert [3]. See Fig. 3 for rotor states. A good description of these states of operation can be found in work by Hansen *et al.* [4].

The angle θ' is the blade pitch angle and θ' is the relative wind angle. If the pitch angle is greater than the relative wind angle, then the rotor acts a propeller and energy is imparted on the fluid particles. This implies that energy has to be supplied to the rotor to maintain a constant rotation speed.

Typical design operation for wind turbines occurs in the windmill state with the wind turbine generating power from the wind. This state is characterised by the blade pitch angle being positive and equal to or less than the relative wind angle. This implies a positive angle of attack and corresponds to an induction factor of $0 < a < 0.5$. Momentum theory is valid in this region. For most practical wind turbine blade designs, the flow induction factor rarely exceeds 0.6 and is close to 0.33 in the operational range for a well-designed blade. For induction factor values of $a > 0.5$, momentum theory is no longer valid and torque and power does not follow the momentum theory curve. This stage represents the rotor of a helicopter.

The vortex ring state is a highly unsteady state characterised by collection of tip vortices around the blade tip in a doughnut shape as a vortex rings; the periodic vortex shedding at the blade tips and reformation of the vortices. The propeller break state happens for an induction parameter of $a > 1$. This state is characterised by the pushing back of flow against the direction of the wind. This state is similar to the powered descent of a helicopter.

3 Available methods for aerodynamic modelling of wind turbines

3.1 Blade element momentum theory

Typical wind turbine performance analysis is done based on BEM theory. BEM is also referred to as Strip theory. The BEM theory is based on the premise that the forces acting on the wind turbine blades are solely responsible for the change in axial momentum of the air passing through the swept area of the blades. BEM analysis is a combination of results from blade element theory and momentum theory. Blade element theory refers to the analysis of forces at a section of the blade, as a function of blade geometry. The blade is split into sections or strips along the length of the blade and each section is analysed separately. The results are combined at the end to provide a total power output for the turbine blade. This usually involves using either 2D CFD methods or experiments to produce look-up tables for airfoil lift and drag data at different angles of attack. A large number of comprehensive computer codes, which use this methodology, are currently maintained by National Renewable energy Laboratory (NREL). These can be found documented by Buhl [6].

Although BEM theory based codes are fast in terms of computational time, quite often good correlations are not obtained against experimental results. In a blind comparison with the NREL Phase VI experimental data taken in the NASA Ames 80 ft by 120 ft wind tunnel, the BEMT power predictions differed from the experimental data by 25% to 175% 'under no-yaw, steady-state, and no-stall conditions'. Results at higher wind speeds in stall were especially disappointing with power predictions ranged from 30% to 275% of measured and blade-bending predictions ranged from 60% to 125% of measured [7]. More recent studies since the original comparison, however, have been able to match the experimental data from the Phase VI or MEXICO [8,9] experiments reasonably well [10]. In BEMT models, the largest source of inaccuracy is often the airfoil data used in design. This can be due to limited airfoil data either at the appropriate Reynolds number or in the stall region.

3.1.1 Effect of finite number of blades

Basic BEM theory based analysis assumes that there are sufficient wind turbine blades such that all fluid particles passing through the rotor disc region interacts with a blade and hence leads to same loss of momentum for all fluid particles. However, with only a finite number of blades in a rotor not all particles will interact with the blade and majority will pass between the blades. This will lead to a momentum loss which is dependent on the position of the fluid particle relative to the blades. This creates an axial velocity that will vary around the blade disc. Hence, axial momentum change in the flow will be determined by the average value and the forces determined by local value of axial induced velocity around the blades.

3.1.2 Tip-loss factor

Tip-loss is another factor that is not included in the simple BEM based analysis. The flow approaching a rotor plane is slowed down at the blade and has to move radially to expand and conserve mass flow rate. Hence, flow is not strictly axial and has a radial component in front of the blades, which arises because of the radial pressure gradient with pressure decreasing radially until it matches atmospheric pressure at the tip. The radial momentum change at any point has to be balanced by an equal and opposite change at a diametrically opposite point in the rotor plane. The maximum radial velocity will be at the rotor tip. However, this kinetic energy associated with the tip radial velocity is not available for energy capture since this does not affect the aerodynamic forces on the blade. If no tip-loss is considered, the optimum axial flow induction factor is uniformly 0.33 over the whole swept rotor. However, in the presence of radial flow at the tip, the value of axial induction factor reduces to zero at the edge of the wake, but is a much higher value near the tip. The tip-loss is usually included in the analysis using a factor called Prandtl's tip loss factor. A mathematical derivation of Prandtl's analysis is beyond the scope of this chapter; however, a closed-form solution of the result can be shown to be

$$f(r) = \frac{2}{\pi} \cos^{-1}\left\{e^{-\pi (R_w - r/d)}\right\}, \tag{17}$$

where $R_w - r$ is the distance measured from the wake edge and d is the distance between successive vortex sheet.

The effect of finite number of blades and the tip-loss effect thus causes a variation of axial flow induction factor in both azimuthally and radially. Hence, the azimuthally averaged value of

$$a(r) = a_b(r) f(r), \tag{18}$$

where $a_b(r)$ is the level of axial flow induction factor that occurs locally at the blade element.

The results from a BEM analyzes can be summarised as

$$\dot{m}_{\text{annulus}} = \rho v_1 \left(1 - a_b(r) f(r)\right) 2\pi r \delta r, \tag{19}$$

$$\overline{(v_1 - v_2)} = 2a_b(r) f(r) v_1, \tag{20}$$

$$\bar{F} = 4\pi r \rho v_1^2 \left(1 - a_b(r) f(r)\right) a_b(r) f(r) \delta r. \tag{21}$$

3.2 BEM theory application

Due to the simplified approach to the power and performance prediction based on 2-D airfoil results, these methods, while being computationally efficient, are usually incapable of accurately modelling three-dimensional effects, tower shadow effects and sweep effects. These effects can have significant effect on power generation and pressure distributions on the turbine blades and will not be captured in BEM methods. However, the simplicity and the lower computational cost make BEM an ideal candidate for design and analysis of wind turbines in the industry. Researchers are trying to increase the accuracy of the combined BEM methods by combining the traditional BEM theory with unsteady potential flow theory and dynamic stall models thus increasing the accuracy in the stall and post stall regime.

Generalised actuator disc method, which is an extension of BEM method, integrated with an Euler or N–S method has been used by several researchers [11–13]. A wide variety of aerodynamic methods that range from actuator-disc models requiring the use of tabulated airfoil data to models based on the solution of the full unsteady 3-D Navier–Stokes equation. The advantage of using generalised actuator-disc models is that they are easy to implement, include physical representation of rotor aerodynamics and achieve accurate results. Their main limitation is that blade geometry is not resolved: when computing the flow past actual wind turbines, the aerodynamic forces acting on the rotor are determined from two-dimensional airfoil characteristics, corrected for three-dimensional effects.

3.3 Navier–Stokes solutions

For many years, the momentum theory, as applied to propellers and combined with a blade-element strip theory approach, has been the most popular model for load and performance predictions for wind turbines [14,15]. The Navier–Stokes method solves the flow-governing equations directly. This allows for correct prediction of flow field around the turbine and behind it without a prior knowledge of the airfoil and section properties. This method also allows for the analysis of all modes of operation of the wind turbine including wind speeds and flow regimes. However, this method is also very computationally expensive.

With regard to wakes, most studies have made a distinct division between the near-wake and far-wake regions; the connection between the two regions is not yet completely understood. The region behind the rotor, of up to three diameters downstream is considered as near wake. Here, the effect of the rotor is predominant. The near-wake region is characterised by an intense turbulence generated by the blades, shear and the degradation of tip vortices transporting a wide range of length scales. The far wake is the region beyond the near wake. The objective of most models in the far-wake region is to evaluate the influence of wind turbines in farms.

Turbulence modelling presents a key issue for predicting the complete wake, from the near-wake region just behind the turbine to the far wake. Within the framework of eddy viscosity turbulence closure modelling, good predictions of the mean and turbulent flow fields rely on reasonable descriptions of the turbulent length scale and velocity scale inside the flow. Detached Eddy simulation methods that solves for a two-equation turbulence model near the blade and a large eddy simulation (LES) model further away from the blade are being used frequently by many researchers to reduce the computational costs associated with a full LES solver.

Le Pape and Lecanu also used a structured multi-block solver to compute the flow over a single blade, in a non-inertial rotating reference frame [16] (Fig. 4). It was reported that an accurate representation of the blade root was necessary for the performance predictions at low wind speeds. The *k*–ω SST Reynolds-averaged

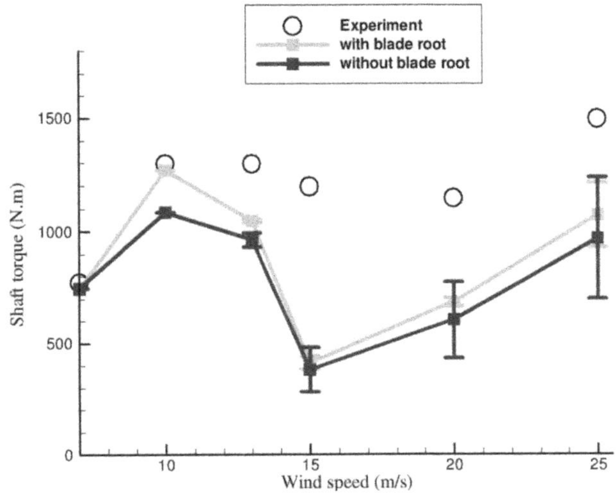

Figure 4: Comparison of measured and computed torque with and without root [16].

Figure 5: Iso-vorticity plot of the instantaneous flow over the turbine for a wind speed of 6.7 m/s.

Navier–Stokes (RANS) turbulence model, which generally performs well for separated flows, was used; however, the torque at moderate to high wind speeds was drastically under predicted due to early stall onset. See Fig. 5 for an iso-vorticity of an instantaneous flow over a turbine with approaching wind speed of 6.7 m/s.

In 2007, Zahle *et al.* applied an incompressible overset method to model rotor–tower interaction on the downwind Sequence B NREL-VI configuration. Due to bluff body vortex shedding by the tower, this was clearly a situation that called for a time-accurate, full Navier–Stokes method. The computations are carried out using the structured grid, incompressible, finite volume flow solver EllipSys3D, which has been extended to include the use of overset grids. The major physics of the blade–wake interaction were captured well. Agreement with experimental data was reduced at higher wind speeds, an effect attributed to the k–ω SST turbulence model. At moderate wind speeds, agreement with experimental data was closest through the mid-span of the blade, where the flow is more nearly two-dimensional. The results show that the rotor has a strong effect on the tower shedding frequency, causing under certain flow conditions vortex lock-in to take place on the upper part of the tower.

3.4 Hybrid methods

The high cost of Navier–Stokes based solvers has driven researchers towards a zonal approach to the solution of wind turbine aerodynamics. This approach combines the robustness and accuracy of the Navier–Stokes solvers and low cost of computation of potential flow solvers. The computational domain is split into three regions:

1. A viscous region around the wind turbine blade where RANS solvers are utilised.

2. A potential flow region surrounding the first region where potential solvers are utilised.
3. A Lagrangian scheme for capturing the tip vortices.

This offers considerable savings in computational work over a full Navier–Stokes analysis.

Xu and Sankar used a hybrid method to study the NREL Phase II and III rotors [17] (see Fig. 6). In this approach, the costly viscous flow equations are solved only in a small viscous flow region surrounding the rotor. The rest of the flow field is modelled using a potential flow methodology. The tip vortices are modelled using a free wake approach, which allows the vortices to deform and interact with

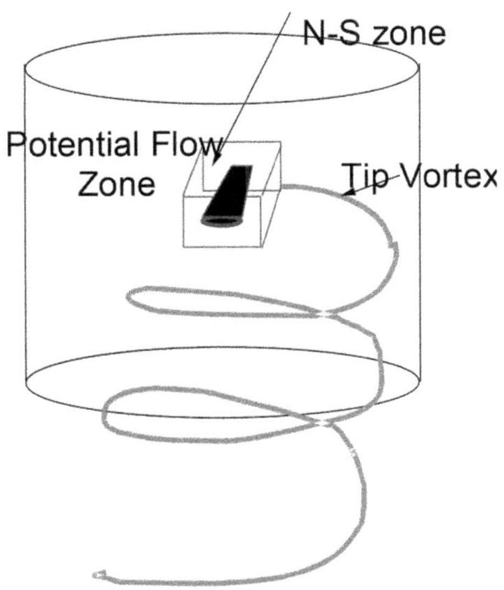

Figure 6: Domain separation for different computational methods [17].

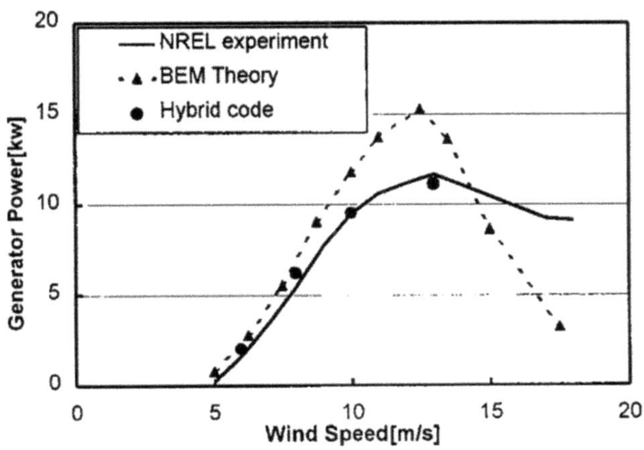

Figure 7: Computed vs. measured power for Phase III NREL rotor.

each other. This hybrid method was also applied to the NREL Phase VI rotor by Benjanirat and Sankar in 2004 [18] (see Fig. 7). It was found that the k–ε turbulence model was not able to predict the flow near the tip with much accuracy there by affecting the rotor torque calculations.

4 Conclusions

This chapter summarises some basics to advanced methods for predicting wind energy characteristics. Some methods introduced are the BEM theory, full computer simulation using RANS method, Prandtl's analysis and advanced turbulence models.

References

[1] Hau, E., *Wind Turbines – Fundamentals, Technologies, Application, Economics*, 2nd edn. Springer-Verlag: Berlin, 2006.

[2] Lanchester, F.W., *The Flying Machine; Two Papers: The Aerofoil and The Screw Propeller*. The Institution of Automobile Engineers: London, 1915.

[3] Glauert, H., *The Elements of Aerofoil and Airscrew Theory*, 2nd edn. Cambridge University Press: Cambridge, Cambridgeshire; New York, 1983.

[4] Hansen, M., *et al.*, State of the art in wind turbine aerodynamics and aeroelasticity. *Progress in Aerospace Sciences*, **42**, pp. 285–330, 2006.

[5] Eggleston, D.M. & Stoddard, F.S., *Wind Turbine Engineering Design*. Van Nostrand Reinhold: New York, 1987.

[6] Buhl, M.L.J., The NWTC design-codes suite: an overview. NREL-NWTC, January 2005.

[7] National Renewable Energy Laboratory (U.S.) and United States Department of Energy. NREL Unsteady Aerodynamics Experiment in the NASA-Ames Wind Tunnel A Comparison of Predictions to Measurements. http://www.osti.gov/servlets/purl/783409/, 2001.

[8] Snel, H., Schepers, J.G. & Montgomerie, B., The MEXICO project (model experiments in controlled conditions): the database and first results of data processing and interpretation. *Journal of Physics: Conference Series*, Paper No. 012014, 2007.

[9] Snel, H., Schepers, J.G. & Siccama, N.B., Mexico Project: the database and results of data processing and interpretation. Presented at the 47th AIAA Aerospace Sciences Meeting, Orlando, FL, 2009.

[10] Martinez, J., Bernabini, L., Probst, O. & Rodriquez, C., An improved BEM model for the power curve prediction of stall-regulated wind turbines. *Wind Energy*, **8**, pp. 385–802, 2005.

[11] Alinot, C. & Masson, C., Aerodynamic simulations of wind turbines operating in atmospheric boundary layer with various thermal stratifications. *ASME Conference Proceedings*, pp. 206–215, 2002.

[12] Mikkelsen, R., Actuator Disc Methods Applied to Wind Turbines, Ph.D. Dissertation, Technical University of Denmark, 2003, MEK-FM-PHD 2003–02 /ISBN 87-7475-296-0
[13] Breton, S.P., *et al.*, Using the actuator surface method to model the three-bladed MESICO wind turbine. Presented at the 48th AIAA Aerospace Sciencing Meeting, Orlando, FL, 2010.
[14] Hansen, M.O.L. & NetLibrary, Inc. (2008). *Aerodynamics of Wind Turbines*, 2nd edn.
[15] Manwell, J.F., McGowan, J.G. & Rogers, A.L., *Wind Energy Explained: Theory, Design and Applications*, 2nd edn. New York, 2009.
[16] Pape, A.L. & Lecanu, J., 3D Navier–Stokes Computations of a Stall-regulated Wind Turbine, *Wind Energy*, **7**, pp. 309–324, 2004.
[17] Xu, G. & Sankar, L.N., Computational study of horizontal axis wind turbines. Presented at the *AIAA*, Paper 99-0042, 1999.
[18] Benjanirat, S. & Sankar, L., Recent improvements to a combined Navier–Stokes full potential methodology for modeling horizontal axis wind turbines. Presented at the *42nd AIAA Aerospace Sciences Meeting and Exhibit*, AIAA-2004-0830, Reno, NV, 2004.

CHAPTER 3

Dynamics-Based Health Monitoring and Control of Wind Turbine Rotors

Douglas E. Adams[1], Scott Dana,[2] Josh Kusnick[2] & Noah Myrent[1]
[1]*Vanderbilt University.*
[2]*Purdue University.*

Abstract

Automatic control and condition monitoring of wind turbines are usually treated as two distinct technical areas because control has a short-term focus, whereas condition monitoring has a long-term focus. In practice, energy capture and condition monitoring actually have interrelated objectives. The structural dynamic behaviour of the rotor system is the link between control and condition monitoring. The aerodynamic loads that govern energy capture are converted by the dynamics of the rotor into drive train loads, which govern many of the failure modes in the turbine. This chapter will discuss two ways in which these rotor structural dynamics can be used to improve decision making in operation and maintenance of wind turbines. A horizontal axis wind turbine test bed is used to experimentally simulate the rotor structural dynamic response to uniform flow as well as horizontal and vertical shear flow across the rotor plane. First, the vibrational characteristics of the rotor system are described in the context of modal analysis where each mode of vibration occurs at a particular frequency with a particular modal deflection shape. These shapes dictate the effectiveness with which a given aerodynamic load couples into the rotor to produce mechanical power in addition to vibrations of the rotor. It is shown using operational modal analysis that changes in yaw and pitch set points change the degree to which the rotor modes of vibration are excited and, consequently, the energy that is captured from the wind and the fatigue loads that act on the turbine. By analysing the changes in rotor response in the flap, edgewise and span directions, the source of power loss is classified so that corrective action can be taken in the form of rotor control. Second, the effects of changes in the wind state on the sensitivity of condition monitoring data to damages in the turbine rotor are discussed. It is shown using modal filters that certain frequencies in the measured response of the wind turbine are sensitive to a change in condition of the rotor in the form of ice accretion or blade root damage for use in detecting that type of damage. By analysing the changes in the modal response amplitudes, the type of damage that is present can be classified.

1 Introduction

Modern utility-scale wind turbines are equipped with several types of sensors for monitoring the wind resource as well as mechanical and electrical variables of interest in the turbine. The performance and reliability of wind turbines are largely governed by the nature of the wind loads that act on the turbine rotor blades. The ability to characterise these loads in real time is advantageous for implementing control algorithms that increase energy capture, which is the primary short-term measure of wind turbine performance. An equally important measure of performance is long-term wind turbine reliability. Reliability is related to the structural integrity of the blade root, low speed shaft, yaw joint and other load bearing components [1]. A more complete awareness of the temporal and spatial variations in the wind loads that act upon the turbine rotor would allow operators to implement advanced control and maintenance strategies. Such information could also provide an understanding of why similar turbines in the same wind farm experience different failure patterns. The methodologies presented in this chapter characterise how the rotor forced response changes due to yaw and pitch set-point errors and how these changes affect the sensitivity of blade measurements to damage mechanisms. This understanding can be applied to improve energy capture while simultaneously facilitating turbine health management. The next two sections review the literature on yaw and pitch control errors and turbine physics of failure and damage detection to establish the need for improved measurement capabilities in the rotor.

1.1 Literature review: performance losses due to yaw and pitch control errors

Utility-scale wind turbines are equipped with anemometers and wind vanes that are located on the nacelle behind the rotor disk to characterise inflow wind conditions. However, this measurement position is not ideal for sensing wind speed due to interference that is created by the blades and nacelle. Furthermore, the wind speed often varies across the rotor disk due to vertical and horizontal wind shear as a result of the atmospheric boundary layer and other phenomena such as wake flow, but a single cup anemometer and wind vane positioned on the nacelle is incapable of measuring these variations. Wind direction data that are gathered by these sensors are used in part to control the yaw position, but these data yield an incomplete perspective of the force distribution on the rotor, which is of most interest for increasing energy capture and monitoring the structural integrity of the turbine. Because of these sensor errors, turbines often suffer from a lack of alignment with the oncoming wind flow; this is known as yaw error. This condition leads to a decrease in energy capture and large fatigue loads on the turbine [2]. Improper yaw alignment has been shown to create a local rotor blade angle of attack near stall. The addition of tower shadow effects and the cyclic separation and reattachment of flow over the rotor blades create large aerodynamic loads beyond static stall values [3]. The source of yaw control errors can be contributed to several different factors such as hysteresis in drive components, improper yaw sensor mesh, development of backlash in yaw position sensors and long time constants between sensor responses and drive action. Wind turbines can experience yaw error routinely due to the dynamic nature of the wind as it continuously changes direction [2]. If winds have high directional variance, yaw error is almost unavoidable especially on short time scales; however, a small increase in yaw response time could significantly improve energy capture by decreasing the turbine response time constant to variations in wind direction.

The development of an active yaw control algorithm has been explored in the literature using the maximum generated power as a means of determining the maximum wind speed direction based on a hill climbing control (HCC) program [4]. This method relies on continuous changes in yaw angle resulting in large time constants for correct perpendicular yaw alignment and an inability to align in non-steady-state wind flow. Other methods utilise costly Doppler LIDAR (Light Detection And Ranging) systems, which apply lasers for three-dimensional mapping of the wind inflow to a turbine. These systems focus on the wind state and do not identify the loads experienced by the rotor blades.

In addition to yaw control errors, wind turbines can also suffer from pitch control errors. The use of active pitch control is common for power regulation. A pitch-controlled wind turbine uses the power output of the generator to determine the necessary pitch action. When power output increases beyond the rated capacity of the turbine's generator, a command is sent from the Supervisory Command And Data Acquisition (SCADA) system to the blade pitch mechanism to immediately turn (pitch) the rotor blades out of the wind. Conversely, as power output drops (i.e. wind speed decreases) the blades are turned back into the wind. This turning action along the longitudinal axis (pitch axis) is controlled by an electronic or hydraulic pitch motor and gear system. Generally, blades are pitched a few degrees every time the wind changes in order to maintain the optimum angle of attack to maximise the power output for all wind speeds. There are also active stall controlled and passive stall controlled wind turbines with different mechanisms for manipulating the blade pitch set point.

Pitch control is the predominant means of control in current utility-scale wind turbines and is highly sensitive to power output. Even a small change in pitch angle (on the order of 2° of less) has a major impact on the performance and health of a wind turbine. The sensitivity to turbine performance is well illustrated by Burton [2]. The sources of pitch error are similar to those for yaw error and include hysteresis in pitch drive components, development of backlash in pitch position sensors and improper installation of the rotor blade in the field.

Power deficits, however, are not solely the result of incorrectly pitched blades. Yaw error, pitch error, rotor or drive train damage, and wind and weather conditions can all lead to reduced power levels. Figure 1 demonstrates this result in a 1 kW horizontal axis wind turbine (described in a later section); note that for normalised power output in the range of 70% and above, the reduction from 100% power can be due to either pitch error or yaw error. To distinguish between the two

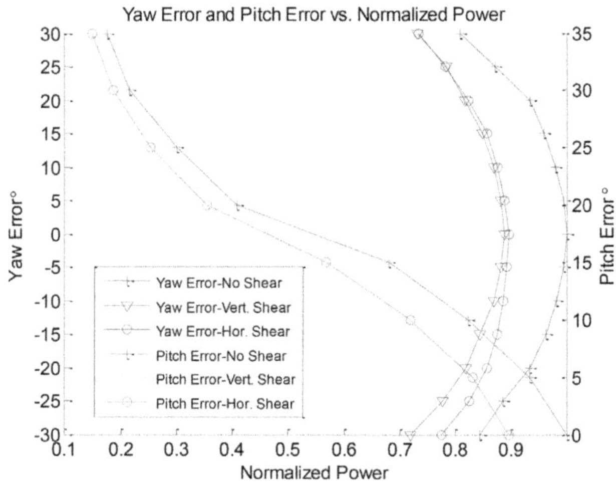

Figure 1: Experimental wind turbine yaw and pitch error vs. normalised generator power output.

it is proposed that the blade dynamic response be analysed rather than relying on point measurements of wind speed, wind direction and generator power for yaw and pitch control. In addition to being able to more precisely determine the wind loading conditions on the blades, structural dynamic analysis can be applied to monitor the health of the turbine components, as described in the next section.

1.2 Literature review: wind turbine damage mechanisms and failures

The premise of this chapter is that short-term (energy based) and long-term (reliability based) measures of performance of wind turbines are interrelated via the rotor response. It has been shown that an increase in turbine power output as wind turbines have grown in size is indeed correlated with a decrease in turbine reliability [5]. A doubling of the rated power of a turbine from 500 kW or less to more than 1,000 kW has doubled the annual rate of reliability incidents. Additionally, the rated power of most utility-scale wind turbines places these machines into the reliability incident-prone category. This failure rate is partly the result of new designs for larger turbines due to the high demand for wind energy. Turbines at a rated power <500 kW have existed for 20 years or more and have experienced multiple design iterations to mature the technology.

Wind turbine blades account for approximately 7% of unforeseen malfunctions or failure, according to Hahn *et al.* [6]. While this percentage is not particularly high compared other subsystems like the gearbox and bearings, turbine blades account for 15%–20% of the total turbine cost. Blade damage is also the most expensive damage to repair and has the longest repair time [7]. Additionally, blade failure leads to secondary component damage in the entire system if prompt repair action is not taken, resulting in greater monetary losses [8]. Blades can become damaged in various ways due to their complex geometry and materials. Turbine blades are generally fibre-reinforced composites with the majority of wind turbine blades being made of glass fibre/epoxy, glass fibre/polyester, wood/epoxy or carbon fibre/epoxy composites [9].

Gearboxes account for approximately 4% of failures according to Hahn *et al.* [6]. It is believed that most gearbox failures do not begin as gear failures or gear-tooth design deficiencies. The observed failures appear to initiate at several specific bearing locations under certain applications, which may later advance into the gear teeth as bearing debris and excess clearances cause surface wear and misalignments [10]. More than 15% of these failures take place in the first five years [11]. These failures occur in spite of the fact that most gearboxes have been designed and developed using the best bearing-design practices available [10]. It is speculated that the primary source of gearbox failure is due to the large bending moments applied at the low-speed shaft from the rotor hub. These loads are produced by the complex rotor dynamics, which are a consequence of wind loading intensity and variability [12].

Another reliability concern for wind turbine rotor blades is blade ice accretion. As the number of wind turbines installed in cold weather climates or regions with cold and wet winters increases, blade ice accretion becomes a major reliability concern [13]. Winter often brings the most favourable wind conditions, but downtime due to blade icing or related damage due to blade icing must be avoided. Without effective ice detection and removal, wind turbines can suffer power reductions as high as 30% per year [14]. In addition to the economic loss, an example of related damage due to blade icing includes increased wind loading experienced by the rotor blade due to a decrease in aerodynamic performance and a rotor imbalance that can severely affect the drive train by means of undesirable torsional loads [13].

An equally important concern of ice accretion is ice throw. As ice accumulates on the rotor blade in operation, shear and centrifugal forces act on the amassed ice, eventually leading to ice throw. It is most critical to detect ice accretion when the turbine is operating in an idle-speed condition. Here, the aerodynamic and centrifugal forces are small, as the rotational speed of the turbine is near zero. This permits the leading edge of the rotor blade to accumulate more ice at the stagnation point of the airfoil. When wind conditions become favourable for power production the rotational speed increases, the angle of attack changes and large masses of ice are shed [14]. According to Cattin *et al.* (2009), ice throw fragments up to 16 in. in length and weighing up to two pounds have been recorded over 300 feet from the nearest wind turbine. Ice throw of this magnitude can pose a risk to civil structures and human life.

To improve the longevity of wind turbines, the sources of reliability shortcomings must be identified. It has been shown that variations in the rotor speed are one of the foremost causes of drive train failure [15]. One of the many sources of these cyclic torsional perturbations is the once per revolution (1 rot^{-1}) speed fluctuation encountered during aerodynamic asymmetry [16]. These torsional oscillations affect not only the gearbox but also they adversely affect the health of blades, bearings, electric generators and all other torsional components. One such source of aerodynamic asymmetries is yaw error, as discussed in the previous section, which occurs when the rotor blades encounter cyclic variations in the blade's angle of attack causing torsional vibrations at 1 rot^{-1} [2].

1.3 Literature review: damage detection methods in rotor blades

Damage detection methods have been applied to wind turbines in offline testing to detect damage of the types mentioned in the previous section. For example, acoustic emission sensors have been used to detect the initiation and propagation of cracks in blades, but the effectiveness of this method requires that the sensor location be relatively close to the damage location [17]. Fibre Bragg grating (FBG) strain gauges have been installed along the entire span of rotor blades to detect cracks. FBGs are immune to electromagnetic interference, and are lightweight; however, they lack the durability of other packaged sensor devices [18]. Digital image correlation (DIC) techniques have been used to determine real-time blade displacement and the global structural integrity by comparing operational video frames with a baseline image [19]. However, these systems are expensive and one system can only monitor a single turbine.

Capacitive accelerometers have also been implemented on utility-scale wind turbines to measure the dynamic response of the blades [12]. The ability to measure both static and dynamic acceleration was shown to provide a means for estimating the deflection of the blade as well as the variations in wind loads acting upon the blade. Modal impact testing of blades is commonly used to estimate the modal frequencies and modal deflection shapes in non-rotating blades [20–22]. Many other researchers have focused on using vibration response measurements from wind turbine rotor blades for the purpose of detecting and monitoring the growth of damage in the blades. For example, various algorithms for detecting damage in wind turbine blades were tested by measuring and processing data using piezoceramic sensor patches [22,23]. However, the effects of operational loading on damage detection were not analysed in these studies.

There are currently no standard solutions on the market for reliable ice detection that can be used as a control input for the turbine's supervisory system. Some measures to prevent icing have been successfully used and deicing methods are in development. However, not all methods operate continuously and there is a need for reliable ice detection to facilitate the activation of the deicing system [24]. Various sensors have been tested but have not performed satisfactorily. Homola *et al.* [24] reviewed 29 methods of ice detection that included direct and indirect methods. The study concluded that only four methods were suitable for ice detection, but with limitations: (1) infrared spectroscopy is limited to monitoring one section of the blade and requires the installation of fibre optic cables in the blade; (2) a flexible resonating diaphragm was shown to be effective but requires installation at multiple points inside the blade; (3) ultrasound has been proven effective at detecting ice on aircraft but has not been implemented in wind turbine blades and is not well suited for retro-fitting and (4) a change in capacitance was also measured using wires mounted in the surface of the blade.

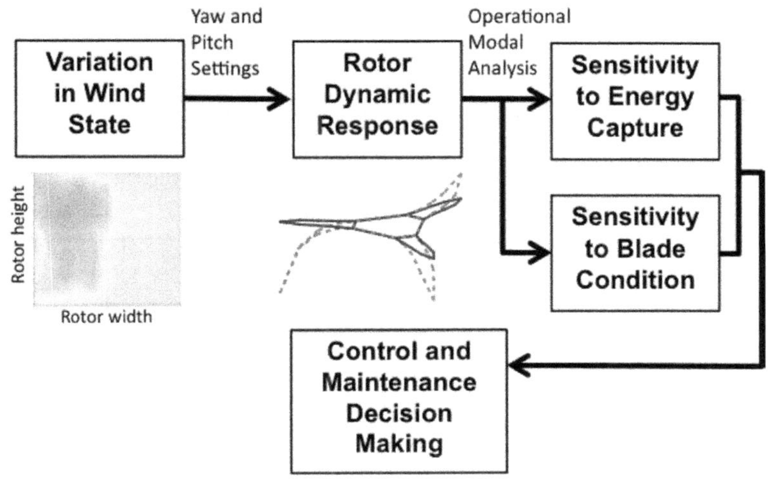

Figure 2: Illustration of technical focus of chapter on the identification and use of rotor structural response for control and maintenance decision making.

1.4 Summary of literature review

Overall, the predictive maintenance of rotor blades and other turbine subsystems can significantly reduce the cost of wind farm maintenance by as much as 50% when compared with the current reactive maintenances approach [25]. This chapter examines the use of dynamic-based sensing within wind turbine rotor blades to facilitate both better maintenance and better control decision-making strategies. The ability to continuously monitor blades during operation has benefits both in terms of the wind farm power output (by identifying changes in the control set points to maximise energy capture) and blade health (by identifying damage that can be repaired early or blade damage that will necessitate blade replacement to avoid the loss of other turbine subsystems). The detection of damage in operational blades is a form of condition monitoring. Operational modal analysis (OMA), which is described in Section 4.2, is one method for acquiring and processing data from the blade to estimate dynamic properties for use in damage detection and control. Figure 2 illustrates the dynamics-based approach that is described and applied in the remainder of this chapter.

2 Experimental Hawt Test Apparatus

2.1 Test equipment

The operational data analysed in this chapter was acquired using the small-scale horizontal axis wind turbine test apparatus that is described in this section. The test apparatus is comprised of the test-bed enclosure and the instrumentation contained within the test-bed. Table 1 lists the major test equipment components and

Table 1: Major test components and specifications of the wind turbine test bed.

Component	Quantity	Specifications
Whisper 100 Wind Turbine	1	SWWP, 2.1 m rotor dia., 900 W at 12.5 m/s
Whisper Power Controller	1	12 V power output, overspeed regulation
DC Tri-axial Accelerometer (Blade mounted)	3	SD, Inc., Model 2460, 80 mV/g, ±50 g pk, 0 to 2 kHz frequency range, 8–32 V DC operation
DC Tri-axial Accelerometer (Nacelle mounted)	1	PCB®, Model 3713B1150G, 40 mV/g, ±50 g pk, 0 to 1 kHz frequency range, 6–30 V DC
Cup Anemometer	1	Omega®, WMS-21 Wind Speed Transducer, Analogue Output
Optical Tachometer	1	Banner Engineering Corp., Mini-Beam® SM312LVQDP, Retro-reflective
Slip Ring	1	Electro-Miniatures Corp., Model 2175, 24 circuit, 1-Amp/Circuit, Variable Speed
cDAQ-9172 Chassis	1	National Instruments™, holds eight I/O modules, USB 2.0-compliant, two 32-bit counter/timer chips
NI® 9234 Input Module	4	4-channel, 24-bit resolution; 102 dB dynamic range; Software-selectable AC/DC coupling; Software-selectable IEPE signal conditioning
NI® 9215 Input Module	1	4-channel, ±10 V 16-bit simultaneous analogue input, 60 V DC CAT I Ch-to-Earth Isolation
DC Power Supply	2	TENMA® 72-6628 Variable Output DC Power Supply

corresponding specifications. Components from this list are referred to by their component name throughout this section.

The centrepiece of the test apparatus is a Whisper 100™ 2.1-m rotor diameter wind turbine, which is manufactured by Southwest Windpower®. This small-scale wind energy converter is assumed to be analogous to a utility-scale wind turbine because both the small-scale and utility-scale turbines use the same physics to convert the kinetic energy of the wind into electricity. To observe the structural dynamics of the rotor, tri-axial Silicon Designs® 2460-050 DC biased accelerometers were mounted in several key locations. Each turbine blade had one of these transducers attached to the low pressure blade using LOCTITE™ 454 industrial instant adhesive at a distance of 14 in. from the blade tip. A non-mechanical means of attaching the sensor, as opposed to a fastener that would have required the blade to be machined, was used to preserve the structural behaviour and integrity of the relatively thin blade. Figure 3 highlights the locations of all sensors and transducers on the wind turbine.

Another DC tri-axial accelerometer (PCB® 3713B1150G) was mounted on the nacelle of the wind turbine. This accelerometer was used to observe the forced response of the turbine outside the rotor blades in a non-rotating reference frame. A cup anemometer was used to collect operational wind speeds during testing. The wind speed transducer was mounted on the nacelle of the turbine behind the rotor. This placement was not ideal for measuring wind speed, but did allow for repeatable measurements and does not interrupt the inflow to the rotor disk. To accurately measure the rotational speed of the wind turbine, an optical tachometer was mounted to the side of the nacelle. Reflective tape was applied to each blade at the rotor hub for use with the tachometer. In addition to the sensors mentioned above, the voltage output of the turbine was also collected from the Whisper Power Controller. This controller rectifies three-phase AC power from the turbine generator to a DC voltage output.

To transmit the DC accelerometer signals from the rotor blades, a slip ring was customised to fit inside a hollow portion of the turbine nacelle, which housed the rotor

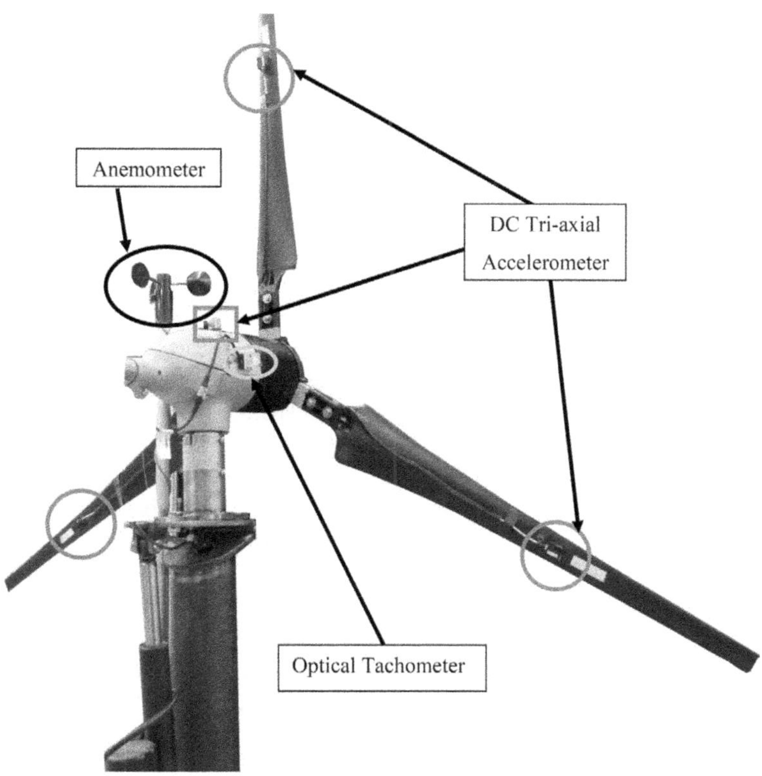

Figure 3: Whisper100 HAWT test apparatus with major components labelled.

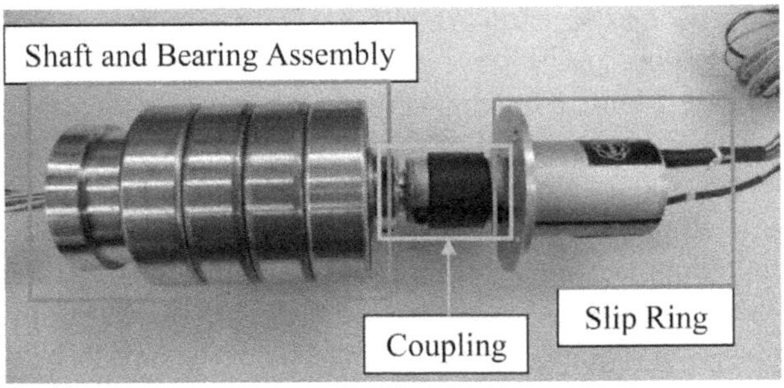

Figure 4: Slip ring attached to rotor shaft and bearing assembly by a flexible coupling.

Figure 5: Yaw set point control fixture. (a) Downwind tower side with degree marks and (b) upwind side of tower.

shaft. Fig. 4 illustrates the rotor shaft and bearing assembly that were attached to the slip ring via a flexible coupling. The slip ring was rigidly fastened to an aluminium collar inside the nacelle. The coupling union was made of neoprene, which offered shock and vibration damping from the heavy rotor shaft to help extend the life of the slip ring. The slip ring enabled 12 channels to be powered and transmitted from the rotating to the non-rotating reference frame: 9 channels were used for the DC tri-axial accelerometers (3 channels for accelerometer per blade) and 1 channel was used for common DC voltage to power the DC tri-axial accelerometers.

2.2 Pitch and yaw control features

The test apparatus was used to set the yaw and pitch to any desired angle. The yaw set point was adjusted by loosening a pair of gear clamps on the turbine tower and then pivoting the nacelle about the yaw axis to the desired degree marker on the yaw joint. The mechanism for setting these angles is pictured in Fig. 5. Once the desired yaw angle was reached, the gear clamps were tightened to hold the yaw

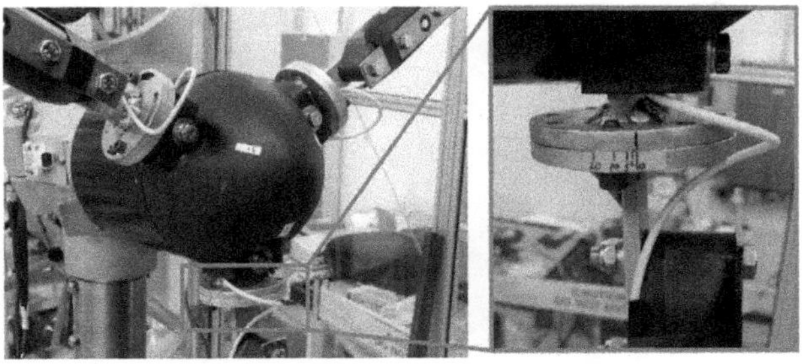

Figure 6: Rotor blade pitch control fixture: (a) rotor blade hub and fixture and (b) detail view of plate interface and degree marks of variable pitch fixture.

position. Yawed flow was tested in this manner. When the yaw degree mark was set to zero, the wind turbine rotor disk was perpendicular to the incoming wind flow resulting in no yaw error. When the turbine's yaw position was adjusted to any degree mark greater than or less than 0°, the turbine experienced the effects of yaw error.

Similarly, the pitch angle of each rotor blade was adjusted by loosening the mounting bolts at the plate interface of the pitch fixture located near the rotor hub and then turning the blade about the pitch axis to the desired angle. Figure 6 shows the installed pitch fixture and a detailed view of the bolted plate interface with degree marks. The pitch fixture was an aluminium bolt-on attachment that was designed to fit the wind turbine. Positive pitch angles were achieved when the leading edge of the blade was turned towards the wind inflow thereby increasing the angle of attack and decreasing the aerodynamic efficiency of the blade. A zero degree pitch angle indicates no pitch error.

2.3 Test-bed enclosure

The test-bed enclosure is pictured in Fig. 7 from the front-side. The Whisper 100 turbine is located inside the enclosure. The test-bed was built on a framework of I-beams (pictured in Fig. 7). The overall footprint of the test bed was approximately 15 × 9.5 ft. The rectangular steel I-beam frame was mounted on four heavy-duty spring hinge casters at each corner to allow the test-bed to be mobile. A floor jack was mounted next to each caster to support the test-bed weight during operational testing. To make the turbine accessible to the user, a plywood floor was installed.

A 9 × 9 ft polycarbonate honeycomb core by Hexcel® with ¼ in. cell size (left side of Fig. 7) was installed at the inlet of the test chamber. The stabiliser core served two functions: (1) develop directional flow down the test chamber and (2) create a quasi-laminar flow state. This core created a nearly uniform airflow over the entire rotor disk that was free of transients due to gusting or lateral velocity components. More discussion of the wind inflow characteristics is included in Section 2.4.2. Airflow was generated at the opposite end of the test frame using four 7.5 hp belt-driven, axial fans (with 48 in blade diameter) that were each capable of

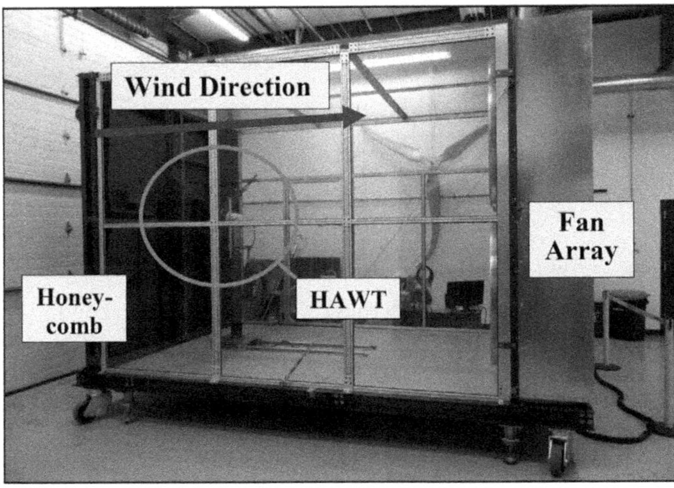

Figure 7: Final design of the 2.1-m rotor diameter HAWT dynamics and control test-bed.

moving 30,000 cubic feet of air per minute. The capacity of the fans was sufficient to generate wind speeds within the operating standards of the wind turbine (4–15 mph). To maintain safety, the exit of each fan was guarded with a heavy gauge sheet metal shroud that was covered with wire mesh to allow for uninterrupted exit flow. The airflow rate was controlled by adjusting the rotational speed of the four fans using a Schneider Electric® Square D® E-Flex™ adjustable speed drive controller that was connected to three-phase 440 V. The motor controller allowed for precise control of the fan speed by 0.1 Hz increments from 0 to 60 Hz (3,600 rpm). To contain generated airflow and to maintain safe operation, an enclosure was fabricated from 2 in. 80/20® Inc. and ¼ in. clear polycarbonate panels.

2.4 Data acquisition hardware and methodology

Operational data were acquired using a National Instruments™ cDAQ-9172 chassis with multiple input modules inserted in the chassis. Three of the four NI 9234 input modules were used to acquire the 12 channels of acceleration data. The fourth NI 9234 input module was used to collect the signal from the optical tachometer. The NI 9215 input module was used to acquire the cup anemometer signal from the digital readout and the voltage output signal of the wind turbine from the power controller. Figure 8 shows the DAQ with input modules, DC power supplies and anemometer digital readout.

The benefit of DC coupled accelerometers, and the reason they were chosen over AC coupled accelerometers, is demonstrated in the differences between the blade responses that are plotted in Fig. 9. The DC-coupled sensor provided accurate readings of both the quasi-static (<5 Hz) and dynamic acceleration components of the blade at the sensor attachment location [20].

Figure 8: DAQ hardware, DC power supplies (2) and cup anemometer digital readout.

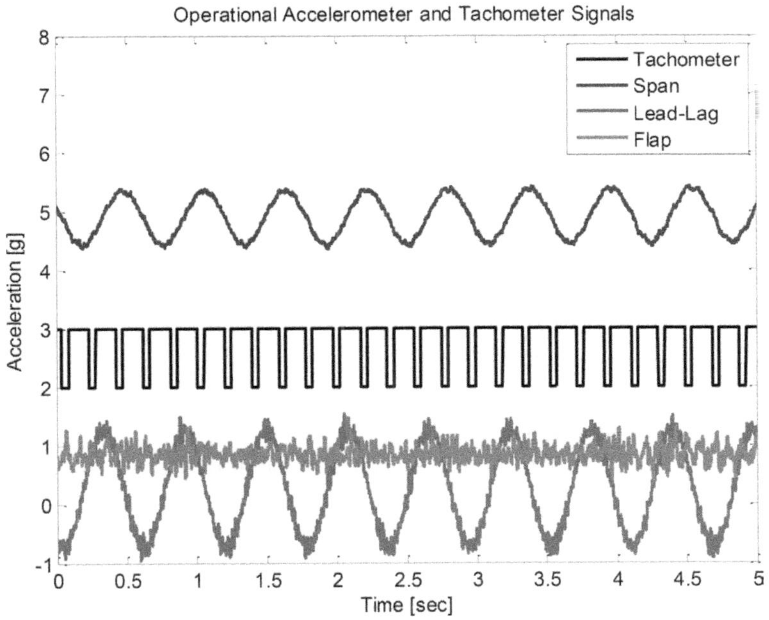

Figure 9: Sample time history from one tri-axial accelerometer and the processed tachometer signal.

2.4.1 Sample data set

A sample forced response time history of approximately eight turbine revolutions is plotted in Fig. 9. The tachometer/keyphasor signal was processed to remove all noise to produce a square wave that oscillates with a magnitude between 0 and 1.

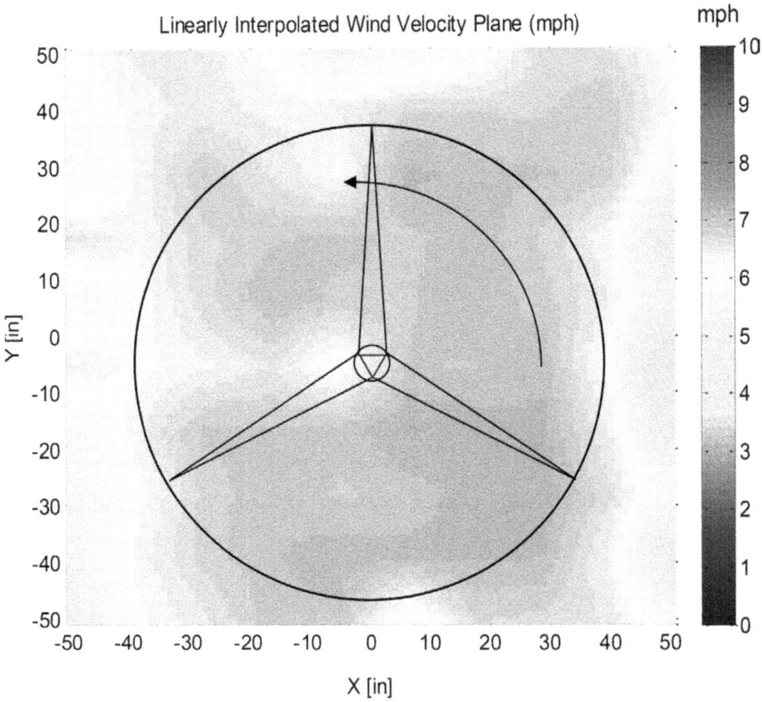

Figure 10: Measured wind profile for the uniform wind condition at 35 Hz fan speed.

In Fig. 9, the accelerometer data are plotted in units of g's (1 g = 9.81 m/s^2). The largest DC offset (signal mean) is observed in the span (x) direction due to the centripetal acceleration along the radial axis. Three contiguous one revolution samples are representative of the data segment lengths that were used in the synchronous averaging procedure.

2.4.2 Wind shear characterisation

To characterise how the rotor forced response of the turbine changed due to yaw and pitch set-point errors and how these changes affected the sensitivity of blade measurements to damage mechanisms, it was important to quantify the forcing function acting on the rotor disk. To do this, wind speed measurements were taken at the inlet of the test-bed enclosure to quantify the wind profile for horizontal and vertical wind shear as well as a uniform wind condition. A 10 × 10 matrix of wind speeds totalling 100 discrete wind speed data points were sampled using the cup anemometer placed 6 in. from the honeycomb. The data were linearly interpolated between the sampled points and a colour bar was added to better highlight the variation in wind speed. The resulting profile for a non-shear wind condition is plotted in Fig. 10. The transparent disk when overlaid on the velocity contour map represents the area of the rotor disk during operation and the arrow indicates the direction of rotation.

To create a vertical and horizontal wind shear condition, screening material was used to shape the wind inflow at the inlet of the test chamber. For the horizontal (side) shear condition, the screening material was used to restrict airflow on the right side enclosure inlet and create a horizontal gradient in wind speed. For the vertical wind shear condition, the screening material was used to restrict airflow on the lower half of the enclosure inlet and create a vertical gradient in wind speed. The resulting wind speed contours for the negative side shear condition and vertical shear condition are plotted in Fig. 11.

2.4.3 Operational testing procedures

A total of 189 tests were conducted, as outlined in Table 2. In each test, the fan speed was held constant at 35 Hz or 2,100 rpm to generate wind speeds of approximately 7 mph. The test duration was 10 min and began when the rotor had reached its steady-state rotational speed. The test data were later segmented into an ensemble of ten, 1-min duration consecutive data sets to allow for the application of statistical analysis, as outlined in Section 4.5. Damage conditions 1, 2 and 3 in Table 2 correspond to no damage, blade ice accretion and a blade

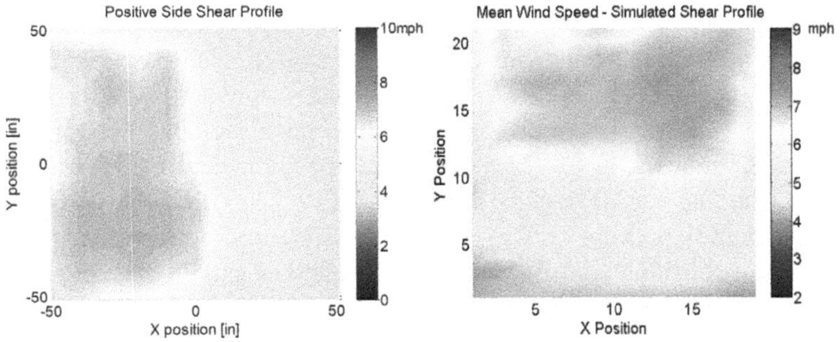

Figure 11: Measured wind profile for the side (horizontal) shear condition and the vertical shear condition at 35 Hz fan speed.

Table 2: Outline of tests conducted.

Test parameter	Wind state	Error angles	Damage condition
Yaw error	Uniform	–30° to 30° in 5° increments	1, 2, 3
Yaw error	Vertical shear	–30° to 30° in 5° increments	1, 2, 3
Yaw error	Horizontal shear	–30° to 30° in 5° increments	1, 2, 3
Pitch error	Uniform	0° to 35° in 5° increments	1, 2, 3
Pitch error	Vertical shear	0° to 35° in 5° increments	1, 2, 3
Pitch error	Horizontal shear	0° to 35° in 5° increments	1, 2, 3

Figure 12: Rubber grommets for simulated root crack.

root crack, respectively. The damage conditions were tested through a range of yaw error angles, pitch error angles and wind states to determine how the blade dynamic response changes due to both variations in the wind state and damage to the rotor. Note that for pitch error testing, only a single blade was pitched away from 0°, representing the case in which (a) a blade was bolted on in the incorrect position during installation or (b) a single blade has developed hysteresis in its pitch drive components or backlash in its pitch position sensors. This blade was the same blade that was damaged during those tests. The pristine, or undamaged, turbine test data served not only as a baseline reference when applying damage detection algorithms but was also used to estimate yaw and pitch errors based on the analysis of the blade acceleration measurements.

To simulate ice accretion, thin gauge sheet metal was cut to match the shape of blade profile and then fixed to the leading edge and high-pressure side of each blade with masking tape. This added material was meant to disrupt the aerodynamic performance of the airfoil as well as to add distributed mass to the blades. The added mass was designed to be 1/20th of the blade mass, which was an amount determined using the dimensions of a Vestas AL40 wind turbine blade and the density of ice to mimic a realistic icing event. However, icing events arise from many complex meteorological events that result in a wide variety of ice formations. The simulated ice accretion in this case does not emulate the surface roughness of ice nor the changes in ice accumulation during operation. A blade root crack was simulated by inserting rubber grommets between the joining plates of the pitch adjustment fixture, thereby decreasing the stiffness in the joint, especially in the flap direction. Figure 12 shows the grommets and the resulting gap between the pitch adjustment plates; this gap was measured each time the pitch was varied during testing to ensure that the stiffness was constant for each pitch angle.

3 Modal Analysis of the Wind Turbine Rotor Blades

3.1 Introduction

Wind turbine rotor blade modes of vibration are excited by wind loads to a greater or lesser degree depending on both the frequency of wind loading and the spatial distribution of that loading. The first step taken to identify changes in the operating state of the wind turbine rotor was to characterise the free dynamic response of the three-bladed rotor system. Knowledge of the free response of the rotor is critical when attempting to monitor the health of the blades because the operational response is a convolution of the blade free response characteristics and the blade forcing function. The mode shapes and resonant frequencies of the rotor dictate the modes of vibration that will be most sensitive to changes in the turbine's operating state (i.e. changes in wind state and/or changes in yaw and pitch set point) and blade damage condition.

3.2 Experimental approach for modal testing

To identify the free response characteristics of the wind turbine rotor system, a modal impact test was performed. A multi-reference modal impact test was conducted using three measurement degrees of freedom (DOF) that were local to each rotor blade's coordinate system. The measurement degrees of freedom are denoted as follows: X: span, Y: lead-lag and Z: flap (flap is measured perpendicular to the blade surface). Figure 13 illustrates the local degrees of freedom of a rotor blade. In total, there were nine reference channels of data, which were acquired using an Agilent E8401A VXI mainframe that was paired with an E1432A module, which sampled at 51.2 kHz. These nine channels consisted of the three DC tri-axial accelerometers as described in Section 2.1. A PCB Piezotronics modally tuned hammer (model 086C01, nominal sensitivity 50 mV/lbf) was used to apply and measure the impulsive forces in the direction perpendicular to the blade at each point that is indicated in Fig. 13 (labelled 'Impact Locations'). The modal test grid consisted of 27 impact points that followed the geometry of the blade. This impact grid offered sufficient spatial resolution to clearly identify and distinguish mode shapes. A threshold force window was applied to the force time history to minimise the noise on the force data outside the applied force pulse. After applying five modal impacts at each location and measuring both the impact force and response time histories corresponding to these impacts, the single–input, multiple-output frequency response functions (FRFs) were estimated for the five averages using the H1 algorithm to minimise the effects of noise on the response measurements [26]. Ordinary coherence functions were also estimated to determine the quality of fit of the FRF models.

3.3 Experimental results from modal impact testing

The multiple-input, multiple-output FRF data were analysed using the Complex Mode Indicator Function, or CMIF [27]. The CMIF is mathematically

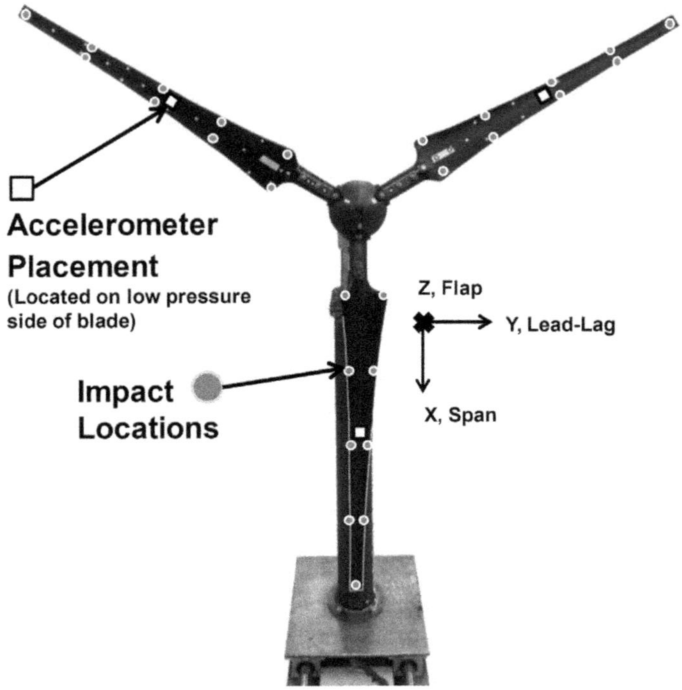

Figure 13: Details of accelerometer placement, input and output measurement degrees of freedom and local blade degrees of freedom on the Whisper 100 wind turbine.

well suited to identify closely spaced modal frequencies by using the singular value decomposition of the normal matrix that was produced using the matrix of FRF data. However, a detailed explanation of the modal and mathematical theory is beyond the scope of this chapter. The results of this computation are illustrated in Fig. 14. The top three spectral lines of this plot were used to identify degenerate modes of vibration with repeated roots and pseudo-repeated roots. At frequencies shared by these three spectral lines, multiple roots are said to exist. One of the more dominant modes was identified at 8.59 Hz for which there were three roots corresponding to this single mode of vibration; consequently, there were three modal deflection shapes associated with this mode of vibration.

Table 3 provides a summary of the damped natural frequencies represented by each peak and multiple peaks found in the CMIF plot. The mode shape descriptions were developed by observing the animated mode shapes that are described in the next section. The mode at 8.59 Hz is a pure first bending mode with a large magnitude of response as seen in Fig. 14.

3.3.1 Modal deflection shapes of wind turbine rotor blades

To better categorise the free response behaviour, the mode shapes were animated and the behaviour of each mode shape was observed. The descriptions provided

Table 3: Summary of turbine modes of vibration identified using CMIF.

ωd (Hz)	Mode shape description
3.13	1st bending – asymmetric bending + slight blade torsion w/ hub rocking
7.03 (2 roots)	1st bending – asymmetric bending w/ one blade in 1st bending and other blades bending w/ hub rocking
7.81 (2 roots)	1st bending – all roots w/ asymmetric blade motion w/ different phase on each root, slight rocking of hub
8.59 (3 roots)	1st bending – asymmetric bending, no rocking of hub. Two blades bend in phase and the third blade is out of phase
13.67	1st bending – asymmetric bending + slight blade torsion w/ hub motion in and out of rotor plane in phase with bend
18.75 (3 roots)	1st bending – asymmetric bending w/ hub rocking. Two blades bend in phase and the third blade is out of phase
20.31	1st bending – pseudo repeated root of 18.75 Hz
22.27 (2 roots)	1st bending – asymmetric bending w/hub motion in and out of rotor plane
24.61	1st bending – asymmetric bending w/blade torsion

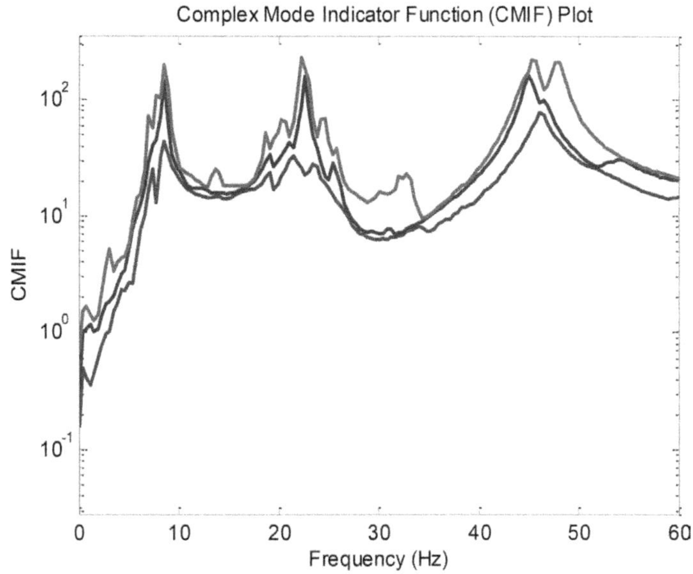

Figure 14: CMIF plot used to identify natural frequencies of the turbine rotor and hub assembly.

in Table 3 are based on the animation. Figure 15 shows the three modal deflection shapes associated with the damped natural frequency near 8.6 Hz (3 roots). These deflection shapes illustrate the phase difference between blades in the modal deflections for the case of a repeated root. Figure 15(a) and (b) exhibits

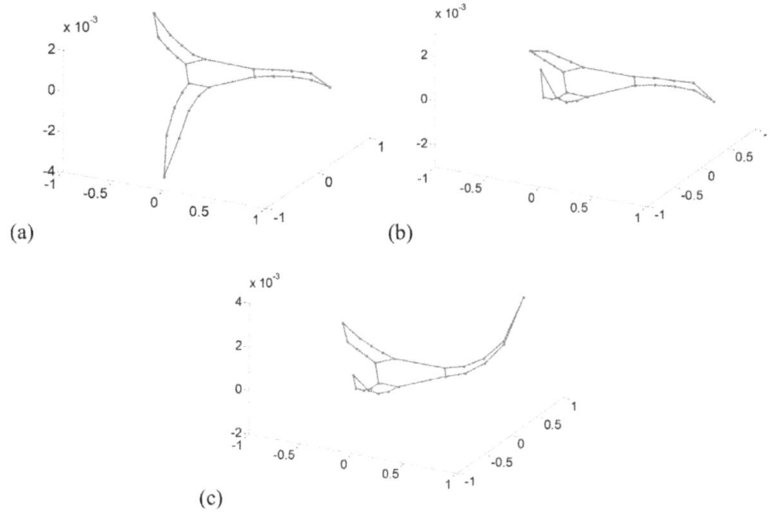

Figure 15: Deflection shapes for modes of vibration near 8.6 Hz showing phase difference between repeated roots. Axes: x, y and deflection amplitude. (a) Asymmetric bending, (b) asymmetric bending of second root and (c) symmetric bending of third root.

first bending with two blades bending in phase and the third blade bending out of phase. It is believed that this mode of vibration and its associated deflection patterns are related to the asymmetric loading that is experienced during operation such as during horizontal or vertical wind shear inflow conditions. Figure 15(c) exhibits symmetric bending in which all three blades move in phase. This mode shape is often called an umbrella mode.

4 Data Analysis Approaches

4.1 Cyclic averaging

Accelerometer measurement data will contain some level of random noise. To improve the accuracy with which the time and frequency data can be analysed, time-synchronous averaging is applied. Blocks of time-sampled data are synchronised using the optical tachometer pulses so that they each begin at the same angular position of the rotor to eliminate the randomness associated with the differences in phase of the blocks of data that are averaged. Components of the signal that are synchronised with the trigger, which is the optical tachometer pulse in this case, are retained, while random noise is averaged out. A block size of three rotations of the rotor is used throughout this analysis so that the averaged time history is long enough to achieve sufficient frequency resolution, which is equal to the inverse of the time history length.

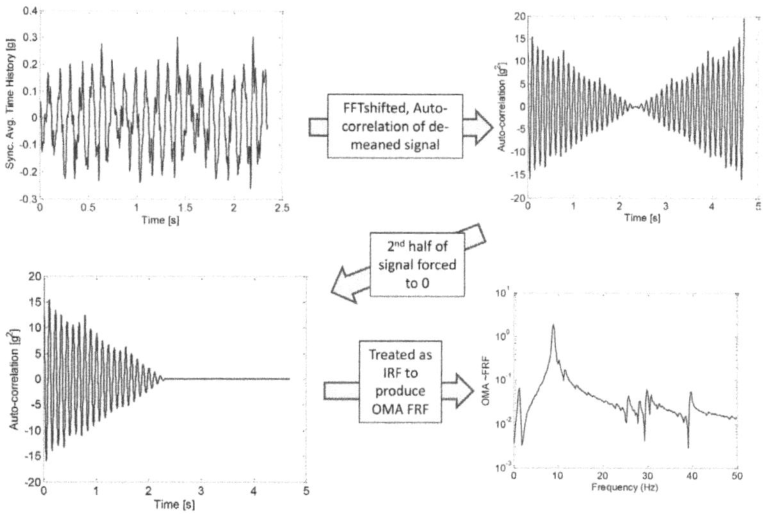

Figure 16: OMA data processing flowchart [30].

4.2 Operational modal analysis (OMA)

The velocity field on the wind plane at the inlet to the rotor disk is experimentally measured as described in Section 2.4.2 for each wind shear condition, but the force distribution along the blades is unknown. To generate FRFs from the accelerometer data that is acquired from the rotor blades, OMA is applied. The two primary assumptions in OMA are: (1) the power spectrum of the input force is broadband and smooth, that is, has no poles or zeroes in the frequency range of interest, and (2) the forcing function is spatially distributed in a uniform manner [28]. Assumption (1) is not particularly applicable to wind-excited structures because the power spectrum of wind is generally dominated by low-frequency components [2]. In general, rotating machinery is self-excited at harmonics of the operating speed [29], and these harmonics are clearly evident in the OMA FRFs that are measured from the turbine being tested in this chapter. Assumption (2) is reasonable for a rotating wind turbine, and the measured wind planes in Figs. 10 and 11 demonstrate the spatial uniformity of the wind speed across the rotor disk inlet.

The OMA procedure consists of first computing the autocorrelation of the synchronously averaged time response for each channel of data. The result is two-sided, containing a positive exponential portion corresponding to the negative poles of the power spectrum, and a decaying exponential portion corresponding to the positive poles [28]. Since both parts of the autocorrelation contain the same information, the positive exponential part is set to zero, essentially zero-padding the time signal. The resulting function is treated as an impulse response function, the discrete Fourier transform of which is the OMA FRF. In dynamic systems, the impulse response and FRFs are related through the Fourier transform in this manner. Figure 16 schematically describes the OMA process.

The discrete Fourier transform is carried out in the order domain, which measures frequency relative to the rotational position rather than absolute time (rotations^{-1} rather than seconds^{-1}). This order domain approach is convenient for analysing the dynamics of rotating machinery for two reasons: (1) as stated previously, these systems are typically self-excited at harmonics of the rotational frequency and (2) variation of the independent parameters throughout testing, such as yaw and pitch error, affects the rotational speed of the rotor; therefore, analysis in the order domain allows for comparisons to be easily made between these unequal-speed data sets.

4.3 Feature extraction

In each experiment conducted, a feature with variable name x is extracted from the processed data to indicate a change in the parameter being studied. In practice, such features are calculated from raw data and then used by the operator to implement control or maintenance decisions. The features that are utilised in this chapter are described in the following subsections.

4.3.1 Yaw error

Due to fluctuations in aerodynamic loading, yaw error is most prevalent in accelerometer data at a once-per-revolution frequency: 1 rot^{-1} in the order domain [30]. The result of integrating the OMA FRF from 0.5 to 1.5 rot^{-1} has been shown to be an indicator of yaw error, even in the presence of horizontal and vertical wind shear [30]. The yaw error data throughout this chapter are analysed using the following procedure:

1. The OMA FRF is calculated for each measurement channel including the flap, lead-lag and span directions on each blade.
2. The OMA FRFs of blade DOF groups (i.e. the flap, lead-lag and span groups) are averaged, resulting in one OMA FRF for each direction associated with the three-blade degrees of freedom.
3. Each of the averaged OMA FRFs is integrated from 0.5 to 1.5 rot^{-1}.
4. The integration results, one for each DOF, are the feature values, x.

4.3.2 Pitch error

Recall that only a single blade underwent pitch error during testing. The blade flap responses couple with the fore-aft acceleration response of the nacelle, and that coupling varies depending on the pitch angle of each blade. To calculate the feature value for pitch error, the following procedure is used:

1. The cross-power spectrum is calculated between each blade's de-meaned, time-synchronously averaged flap acceleration response and the fore-aft acceleration response measured on the nacelle housing. The discrete Fourier transform is performed in the order domain.
2. The peak in each cross-power spectrum magnitude closest to 1 rot^{-1} is found.

3. The pitched blade has a higher magnitude of response in the cross-power spectrum at 1 rot^{-1} than the other two blades. The feature is the sum of the difference of the peak value in the pitched blade's cross-power spectrum at 1 rot^{-1} with the other two blades' peak values, normalised by the sum of the non-pitched blade's peak response. This normalisation accounts for relative magnitude differences due to reduced rotor speed in the high pitch-angle tests. Mathematically, the feature value, x, is:

$$x = \frac{\left[\left(p_3 - p_1\right) + \left(p_3 - p_2\right)\right]}{p_1 + p_2}, \tag{1}$$

where p_3 is the peak in pitched blade's cross-power spectrum magnitude at 1 rot^{-1}, p_1 the peak in non-pitched blade one's cross-power spectrum magnitude at 1 rot^{-1}, p_2 the peak in non-pitched blade two's cross-power spectrum magnitude at 1 rot^{-1}.

4.3.3 Blade damage detection

For damage detection, the time-synchronously averaged operational response data are analysed in the frequency domain. The averaged linear spectra are calculated using the discrete Fourier transform. Figure 17(a) shows an example of the linear spectra magnitude for the case of blade ice accretion and undamaged blades. Many of the peaks in the linear spectra occur at frequencies that correspond to the modes of vibration listed in Table 1. The information gleaned in the modal analysis of the rotor blades helps develop an understanding of which modes are excited during operation. Mode shape analysis is used to identify frequencies in the operational data that correspond to asymmetric bending modes such as modes at 3.1 and 8.6 Hz, but any frequency with a known mode of interest can be evaluated. Trends in the dynamic response at these frequencies can be exploited to reveal a change in the rotor blade condition and how the ability to observe that condition varies with yaw angle.

To evaluate these trends, the mode (frequency) of interest selected in the processed operating frequency spectra and the maximum value of the magnitude of the DC acceleration is recorded for a single DOF for each blade within the frequency band. This procedure is carried out for each minute of the 10-min data set and the 10 values are averaged to provide one value for each blade at each yaw angle. The figures below illustrate the frequency banding process, where the frequency of interest is 3.1 Hz. The linear spectra are then bound from 2 to 4 Hz and the maximum acceleration value is found for each blade on this interval.

The magnitude of the acceleration for each blade is used to compare the change in response due to a damage condition and a yaw angle in various wind regimes. In the case of blade root damage, the responses of the blades are compared by finding the difference in response between blade pairs. In this study, Blade 3 is made to experience a reduction in stiffness at the root boundary condition. The reduction in stiffness should cause a change in operational response of this blade and is measured against the healthy blades; Blades 1 and 2. For this damage case, the

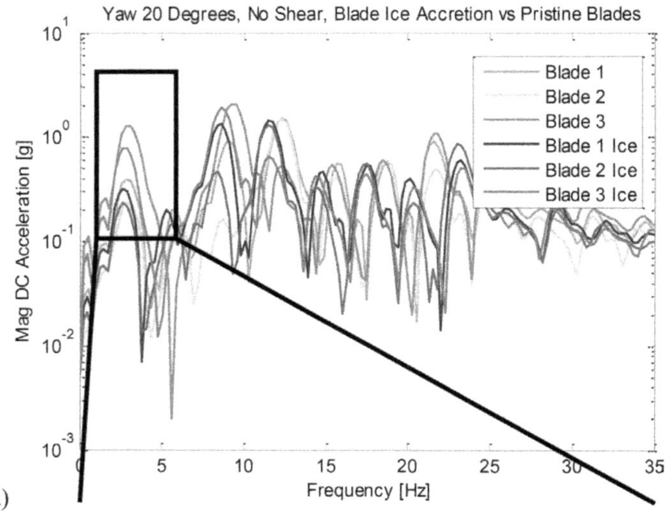

(a)

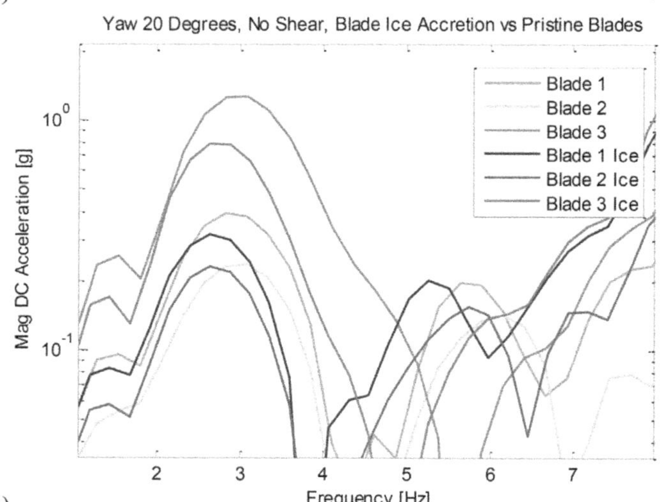

(b)

Figure 17: Example of frequency bounding process. (a) Linear spectra of blade ice accretion and pristine blades for Flap DOF at 20° yaw angle. (b) Detailed view at frequency of interest.

difference in response between Blades 1 and 3 should, therefore, be greater than the difference between Blades 1 and 2.

For the case of ice accretion, the magnitude of acceleration is again used to track changes in operating response of each blade at a particular frequency of interest. Here, the magnitude of the response when ice is present on the rotor blades is compared with the magnitude of the response for un-iced blades by calculating the percent error between the historical baseline response and the iced blade response.

4.4 Statistical modelling

Ten 1-min duration tests are conducted for each test scenario, allowing for the calculation of 10 feature values for each test condition. Each test condition is defined in terms of the corresponding yaw or pitch error, blade damage/condition and wind shear profile. Statistical analysis is applied using hypothesis testing to determine how variations in the wind load affect the sensitivity to yaw or pitch error and the sensitivity to rotor damage/condition. The feature, x, is normalised to obtain its standard score according to

$$\frac{x-\mu}{\sigma}. \tag{2}$$

where μ and σ are the estimated mean and standard deviation of x, respectively. The standard score is how many standard deviations x is above or below its mean. Because eqn (3) is normalised by σ, changes in x are more readily detected when the feature has little variance for a given operating condition. If x is assumed to be equal to $\mu \pm 3\sigma$, then the level of change in x from the mean value corresponds to a 99% confidence interval, that is, the analyst is 99% certain that the feature has undergone a biased change even in the presence of natural variations in x [1]. When this value of x is substituted into eqn (1) and then normalised, the following result is obtained:

$$\left|\frac{x}{\mu}-1\right| = \frac{3\sigma}{\mu}, \tag{3}$$

expressing how much of a shift in x is required to detect a statistically significant change in the feature mean with 99% confidence. For example, if x in the flap direction for yaw error exhibits a σ/μ of 0.02, then a 6% change in x (3*0.02) is required to achieve 99% confidence that a change due to yaw error has occurred. Therefore, the result of eqn (2) represents a measurement of the sensitivity of x: if it is small, then the sensitivity is high, and that is desirable in terms of condition monitoring because it means that, even amidst variations in wind loading and other factors, a small change in x indicates a significant change in the feature.

5 Experimental Results and Discussion

5.1 Yaw error detection

The yaw error feature is calculated in each of the blade degrees of freedom for the three different wind conditions, as described in Section 4.4.1. These feature values are first plotted versus the yaw error angle to produce one curve for each wind condition and rotor blade measurement DOF in Figs. 18–20. Each of these figures has an accompanying sensitivity curve, which is a plot of the percent change in the feature that is required to detect yaw error with 99% confidence, which is

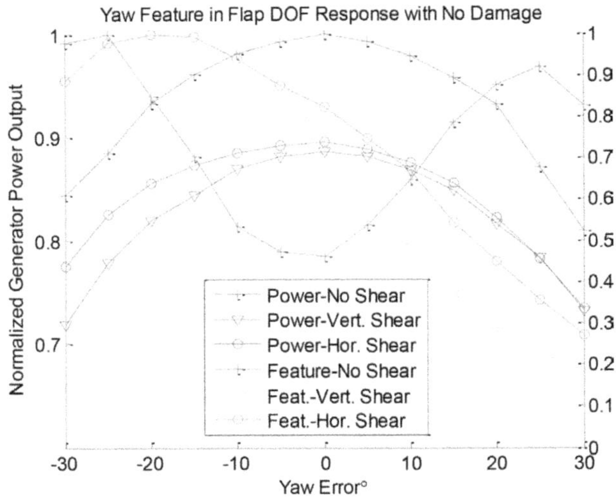

Figure 18: Normalised yaw feature and power vs. yaw error angle for flap DOF.

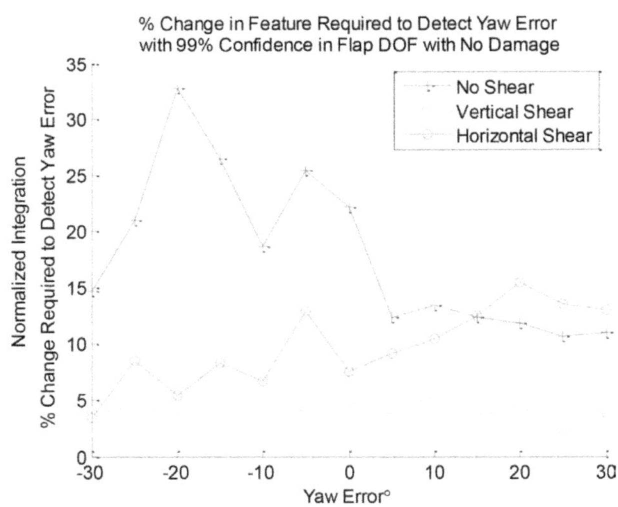

Figure 19: Normalised yaw feature and power vs. yaw error angle for lead-lag DOF.

equivalent to $3\sigma x/\mu x$, as calculated in Section 4.4. A low value on these curves indicates a high sensitivity of the feature at that particular yaw error and wind plane shear condition. For instance, in Fig. 21, the value on the vertical shear sensitivity curve at +10° yaw error is 5%, while the value on the no-shear sensitivity curve is 15%. Thus, it can be said that under a vertically sheared wind profile, only a 5% change in the feature, x, from its mean value is required to detect a yaw error of +10°, but under a uniform wind velocity distribution, a 15% change in the mean value of the feature is required to detect the same +10° yaw error.

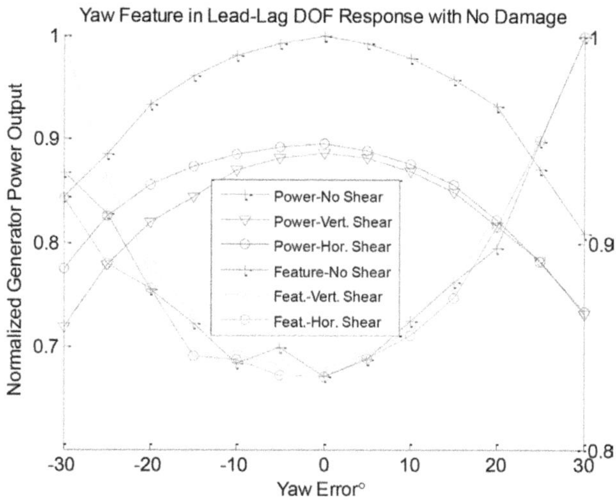

Figure 20: Normalised yaw feature and power vs. yaw error angle for span DOF.

5.1.1 Yaw error feature: flap DOF

Several important findings in the presence of yaw error can be extracted from the plots of the flap measurement DOF response and sensitivity seen in Figs.18 and 21.

First, as expected, the power extracted from the wind in the test cases involving wind shear is lower than the corresponding case involving no wind shear (refer to Fig. 18). Generator power is also reduced due to yaw error.

Secondly, Fig. 18 shows that the feature is symmetric about 0° yaw error for the uniform (no-shear) wind condition, whereas the vertical and horizontal wind shears produce asymmetric curves. Asymmetry can be helpful from a controls standpoint because each yaw error can then be associated with a unique feature value. For instance, if the wind turbine is in a uniform flow and the normalised flap response feature value is 0.5, the yaw error could be either +5° or 10°. If, however, the wind profile is characterised by vertical shear, that same normalised feature value of 0.5 would indicate approximately a +8° yaw error.

Thirdly, it is observed that wind shear does not hinder but rather enhances the ability to detect yaw error. This increase in sensitivity in the presence of wind shear is clearly evident in Fig. 21 because the wind shear curves are much smaller in magnitude and, therefore, much larger in sensitivity as explained previously. The reason why wind shear does not interfere with the feature calculation is that it tends to affect the 3 rot^{-1} blade dynamics [30]; therefore, it does not affect the integration of the OMA FRF over the 0.5–1.5 rot^{-1} range.

Fourthly, under horizontal wind shear the sensitivity decreases (the % change in x required to detect a change in the feature value with 99% confidence increases) as the turbine is yawed into the side of the inlet plane with higher velocity wind (positive yaw error). Horizontal wind shear can be particularly prevalent in wind farms due to the presence of wake flows from upstream turbine rotors, which produce velocity deficits on downstream turbine rotors.

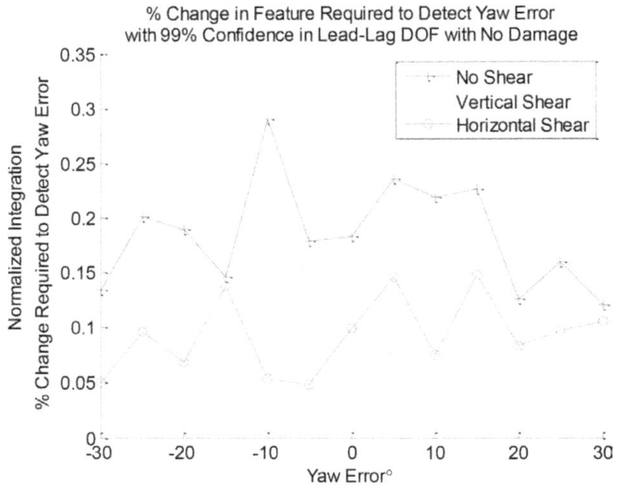

Figure 21: Sensitivity of yaw feature vs. yaw error angle for flap DOF.

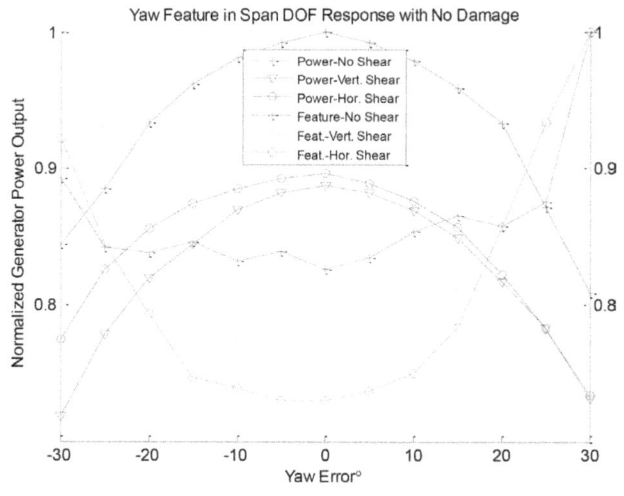

Figure 22: Sensitivity of yaw feature vs. yaw error angle for lead-lag DOF.

5.1.2 Yaw error feature: lead-lag DOF

Figure 19 shows that in the lead-lag DOF, the feature is nearly symmetric about 0° yaw error, regardless of wind shear condition. Furthermore, Fig. 22 indicates that the feature sensitivity is two orders of magnitude higher in the lead-lag direction than in the flap direction of Fig. 21, likely because the blade undergoes only relatively small deflection in the stiffer lead-lag direction, and thus experiences less variation due to factors other than yaw error. Therefore, not only is the standard deviation of the lead-lag measurements lower, but also a higher proportion of the variation in this direction is due to yaw error compared with the flap direction.

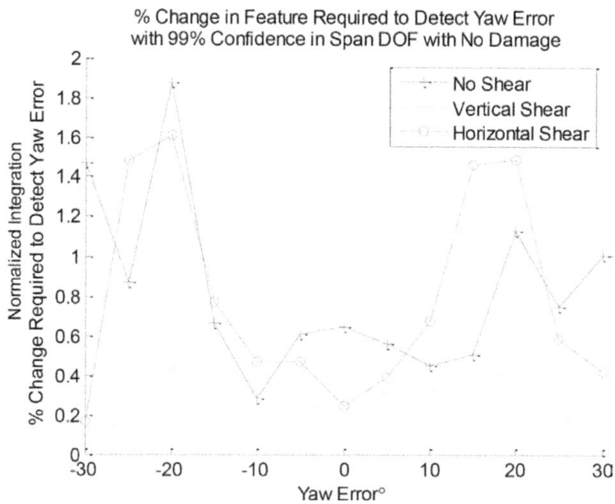

Figure 23: Sensitivity of yaw feature vs. yaw error angle for span DOF.

5.1.3 Yaw error feature: span DOF

The span DOF exhibits similar feature curves as the lead-lag DOF in the vertical and horizontal shear conditions, but the uniform flow condition produces a nearly-flat feature curve between ±20°, which is not desirable because it means that a wide range of yaw errors can have nearly the same feature value. Furthermore, while the feature curves are similar for the vertical and horizontal shear cases, Fig. 23 shows that the feature for span is more sensitive in vertical shear, especially for larger yaw errors.

5.2 Pitch error detection

The method described in Section 4.4.2 was effective at detecting pitch error in the cases of uniform and vertical wind shear conditions, as shown in Fig. 24. Interestingly, the two feature curves are nearly identical, which can be beneficial in terms of control because the method will work consistently regardless of whether there is uniform or vertically sheared wind flow. Under the horizontal wind shear condition, the trend followed closely for pitch errors of 15° or higher, but was not as reliable for pitch errors of 0° and 5°, at which the feature value became negative. One possible explanation for this different behaviour in horizontal wind shear is that due to the severe wind profile that was used, the cross-power spectra at the smaller pitch angles were dominated by the 3 rot^{-1} frequency component, so the blade-to-blade differences in the dynamics at 1/rot^{-1} were not as pronounced.

Note that the other blade DOFs, lead-lag and span, did not exhibit any significant trends in their cross power spectra with any of the nacelle DOFs and therefore were not effective in pitch error detection using this method.

The sensitivity of the pitch feature was very low for 0° and 5°, especially in the vertical and horizontal wind shear cases (see Fig. 25), meaning that a large change in the mean value of the feature is required to detect low pitch errors. This occurred

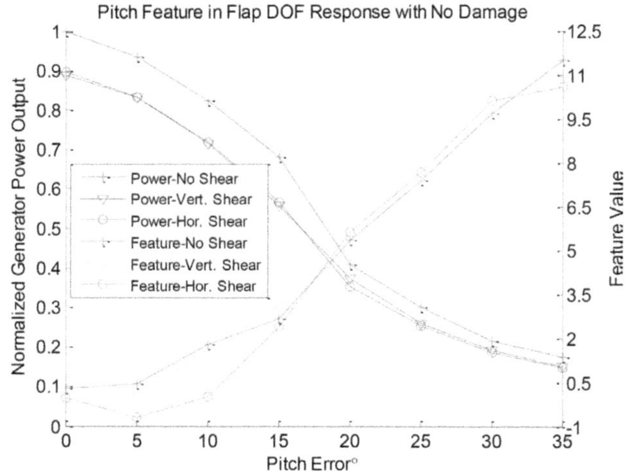

Figure 24: Feature value and power curves for uniform and vertical shear wind conditions.

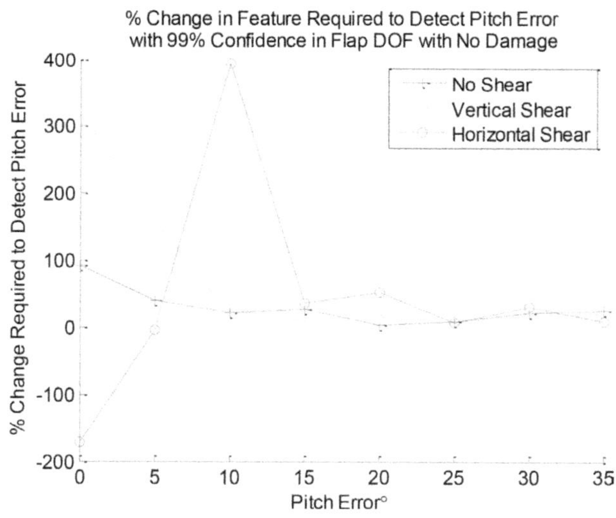

Figure 25: Pitch error sensitivity.

because for low pitch errors, the low mean value of the feature, which is near zero at zero pitch error, is on the order of the standard deviation. Figure 26 shows that the sensitivity of this measurement above 10 degrees pitch error is fairly high.

5.3 Ice accretion detection

By using the approach outlined in Section 4.4.3, the change in response due to blade ice accretion was readily identified. The frequency of most interest for damage detection was found to be 3.1 Hz; therefore, all damage results correspond to the change in the magnitude of acceleration in the 2–4 Hz frequency band of the

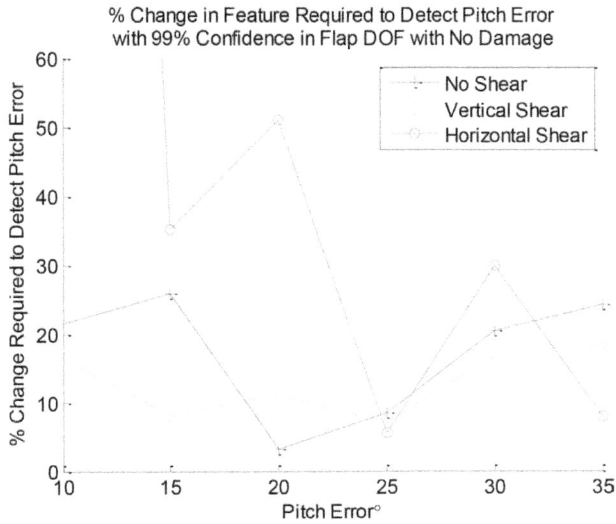

Figure 26: Pitch error sensitivity (10° to 35°).

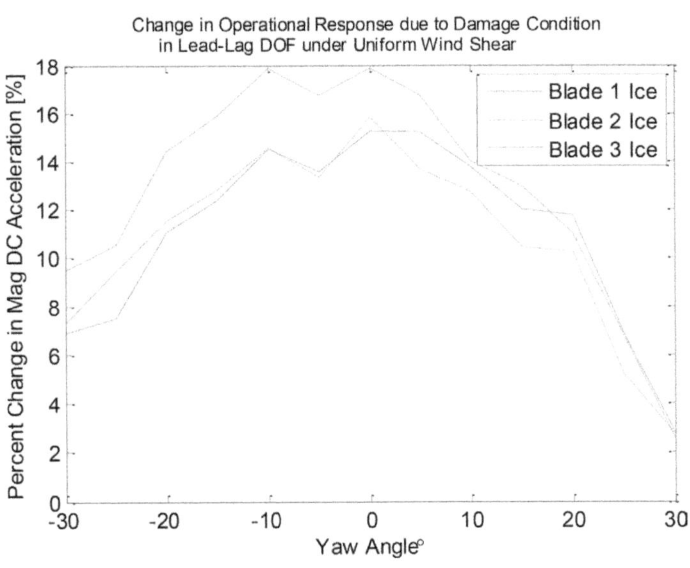

Figure 27: Edge-wise percent change in the magnitude of acceleration for each blade vs. yaw angle at 1 rot^{-1} with ice accretion in uniform wind flow.

linear spectra. In the order domain, this frequency corresponds to 1 rot^{-1} response of the turbine rotor-dynamics. Additionally, the lead-lag DOF revealed the greatest sensitivity to changes in the rotor-dynamics due to ice accretion.

5.3.1 Detection of ice accretion in the presence of yaw error

Figure 27 shows the percent change in the magnitude of response in the edge-wise (lead-lag) direction for each blade plotted vs. yaw angle. This plot reveals that ice accretion causes an increase in the edge-wise response in the case of uniform wind flow for all yaw angles. The response of Blade 3 is slightly higher but the overall change in response of each blade is appreciable. For low yaw angles, the response of each blade has increased by approximately 15% to 18%. As yaw error increases the percent change in response magnitude for each blade decreases, making the ability to detect ice accretion in the presence of severe yaw error more challenging.

A similar trend is observed when operating in vertical shear flow. Figure 28 illustrates that simulated ice accretion causes the edge-wise response to increase significantly for all three blades. This figure also reveals a large measure of symmetry about 0° yaw position. Similarly, the same increase in response is observed when operating in horizontal shear flow, as illustrated in Fig. 29. These results suggest that the damage condition can be identified regardless of the wind profile. This result is important for the application of utility-scale wind turbines where vertical and horizontal wind shear conditions can be prominent in wind farms.

5.3.2 Detection of ice accretion in the presence of pitch error

For the case of pitch error, similar results and trends observed in yaw error were revealed. Figure 30 shows the percent change in the magnitude of response in the edge-wise (lead-lag) direction for each blade plotted vs. pitch angle. This plot reveals that ice accretion causes a significant increase in the edge-wise response of each blade across all yaw angles in the case of uniform wind flow. The response of Blade 2 is slightly lower than Blades 1 and 3; however, the overall percent change in the response of each blade exceeded 35% at low pitch angles from 0° to 5°. As pitch error increases the percent change in response magnitude for each blade decreases. For 35° pitch error, the percent change in the magnitude of response for all blades coalesces near 26%. This is a desirable result that suggested ice

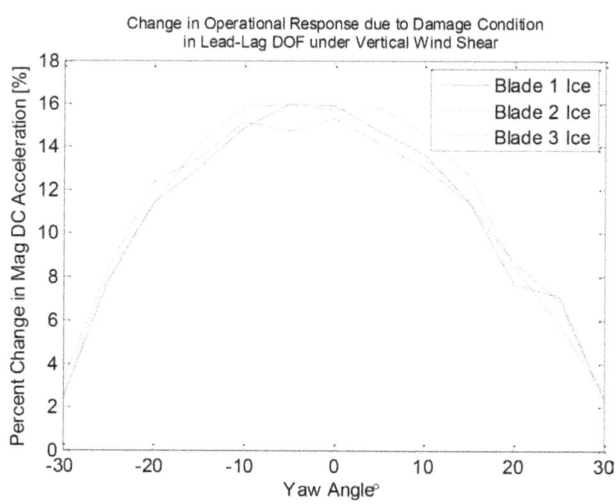

Figure 28: Edge-wise percent change in the magnitude of acceleration for each blade vs. yaw angle with ice accretion in vertical shear flow.

accretion can be detected regardless of the pitch error angle when the wind turbine is operating in uniform wind flow.

For the case of a turbine operating in vertical and horizontal shear regimes, the trends observed in Fig. 30 are similar. Figures 31 and 32 show the percent change in the magnitude of response in the edge-wise (lead-lag) direction for each blade

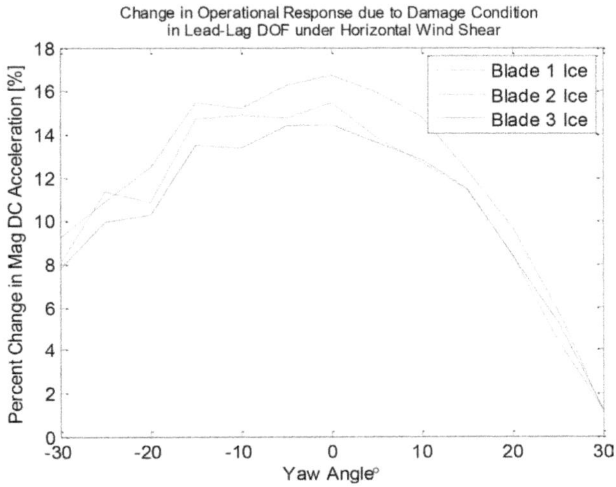

Figure 29: Edge-wise percent change in the magnitude of acceleration for each blade vs. yaw angle with ice accretion in horizontal shear flow.

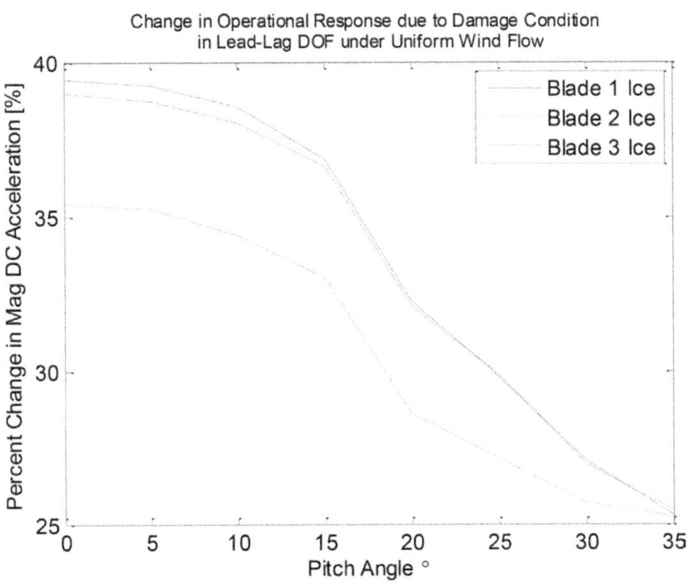

Figure 30: Edge-wise percent change in the magnitude of acceleration for each blade vs. pitch angle at 1 rot^{-1} with ice accretion in uniform wind flow.

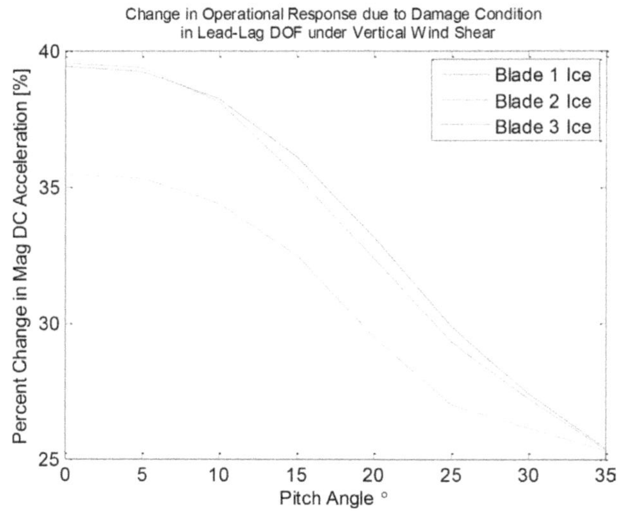

Figure 31: Edge-wise percent change in the magnitude of acceleration for each blade vs. pitch angle with ice accretion in vertical shear flow.

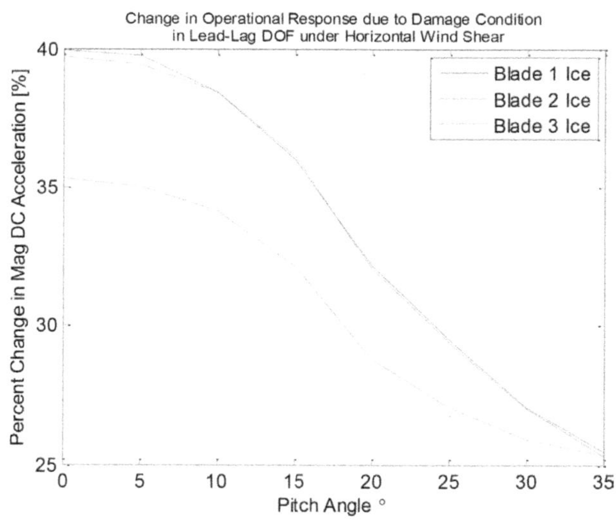

Figure 32: Edge-wise percent change in the magnitude of acceleration for each blade vs. pitch angle with ice accretion in horizontal shear flow.

plotted vs. pitch angle when operating in vertical shear and horizontal shear, respectively. Again, the overall percent change in the response of each blade exceeded 35% at low pitch angles from 0° to 5° and for increasing pitch error the percent change in the magnitude of response of all blades coalesce near 26%. As with detection of ice accretion in the presence of yaw error, these results suggest that the damage condition can be identified regardless of the wind profile or pitch error.

5.4 Damage detection

In the case of blade root damage, the method outlined in Section 4.4.3 was applied and revealed that the flap DOF near 7 Hz emphasised the change in response due to the damage condition. In the order domain, this frequency corresponds to 2 rot^{-1} dynamics of the turbine. Recalling Fig. 11, the rotor experiences a 2 rot^{-1} oscillation in wind speed as it moves through one full rotation. At this rotational frequency, the reduced stiffness in the root of the blade is most sensitive to the 2 rot^{-1} fluctuations in the flap DOF; therefore, all damage results correspond to the change in the magnitude of acceleration between blades at this order.

5.4.1 Root damage detection in the presence of yaw error

Figure 33 shows the change in blade-to-blade response ratio for varying yaw error when operating in a uniform wind flow. The damaged blade (Blade 3) causes the blue and green curves to exhibit the largest changes, whereas there is negligible change near zero yaw error for the undamaged blade-to-blade response ratio (red curve). The undamaged blade-to-blade response ratio exhibits symmetry about zero yaw with a slight increase in response due to the change in yaw position near 10° and 15°. Figure 34 shows a similar trend for the blade-to-blade response ratio when operating under vertical wind shear. In this wind regime the 2 rot^{-1} dynamics are more pronounced due to the shear profile. Again, the blue and green curves exhibit the largest changes whereas there is negligible change for the undamaged

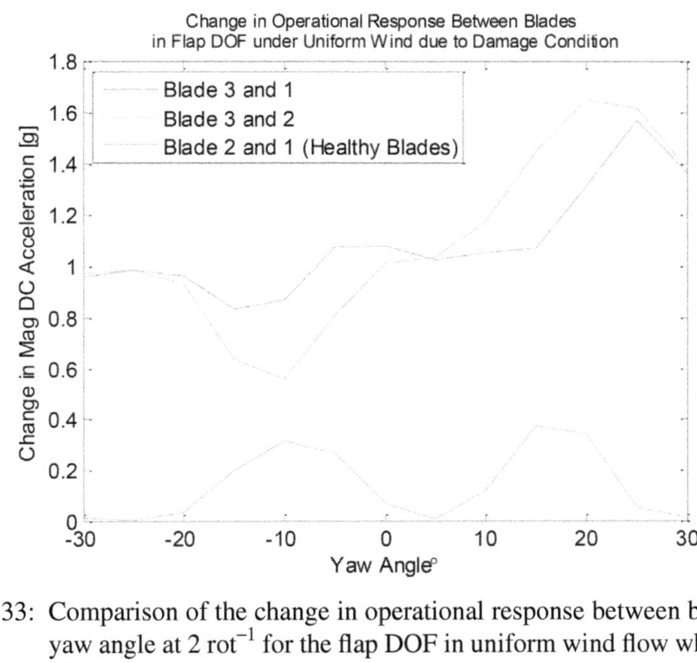

Figure 33: Comparison of the change in operational response between blades vs. yaw angle at 2 rot^{-1} for the flap DOF in uniform wind flow when damage is present in a single rotor blade (Blade 3).

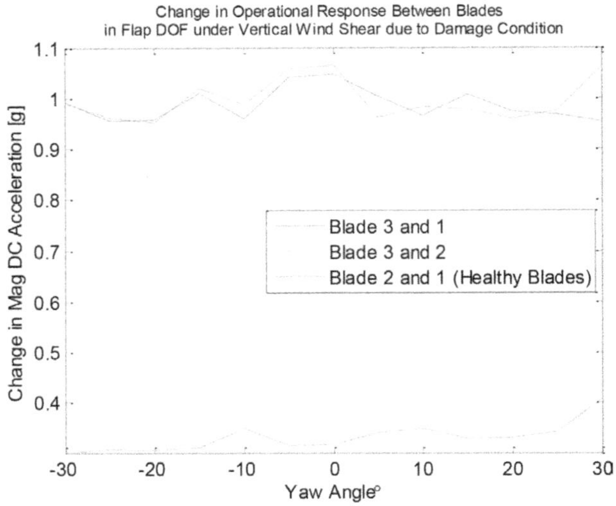

Figure 34: Comparison of the change in operational response between blades vs. yaw angle for the flap DOF in vertical shear wind flow when damage is present in a single blade (Blade 3).

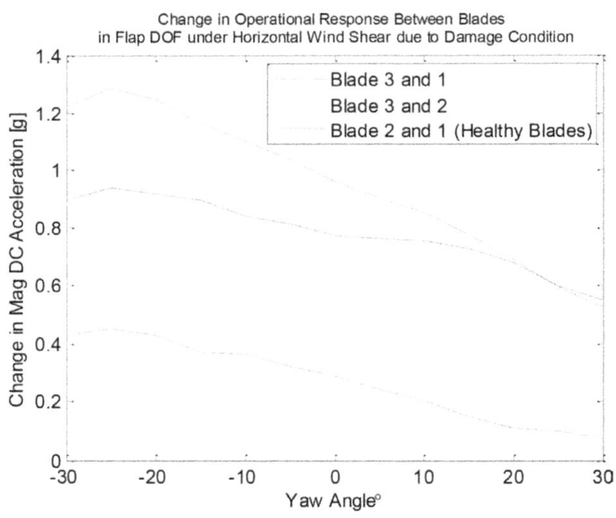

Figure 35: Comparison of the change in operational response between blades vs. yaw angle for the flap DOF in horizontal shear flow when damage is present in a single blade (Blade 3).

blade-to-blade response ratio (red curve). Similar traits are observed in Fig. 35 for a turbine operating under horizontal wind shear. However, the curves now increase for negative yaw error, and the undamaged blade-to-blade response (red curve) has lost the symmetry about zero yaw, but does maintain an appreciably smaller magnitude when compared with the blade-to-blade response ratio with damage present. Recalling Fig. 11, at negative yaw error for horizontal wind shear the rotor plane is oriented in an increased wind flow and therefore experiences an increase in response for negative yaw error. The increased blade-to-blade response ratio for negative yaw error is an artefact of the horizontal wind shear and yaw position combined. Regardless, the blade-to-blade response ratio method identifies root damage in the presence of yaw error. The advantage of this method is that no historical baseline response data are needed to make these comparisons to determine if damage is present.

5.4.2 Root damage detection in the presence of pitch error

As with yaw error, similar trends are observed in the blade-to-blade response ratios to identify damage in the presence of changing pitch angle. Upon first inspection of the blade-to-blade response ratios for each wind regime an obvious trend emerged; the change in response amplitude for the blade pairs with the damaged blade (Blade 3) increased significantly beyond a 15° pitch angle. To understand if this trend was due to the damage condition of Blade 3 or due to the change in pitch of Blade 3, the blade-to-blade response ratios were plotted for the case of no damage and only the pitch angle of Blade 3 was altered. Figure 36 shows the results of this plot. The figure reveals that for pitch angles from 0° to 15° the change in the blade-to-blade response ratios for all blade pairs is negligible, that is, damage is not present. However, for pitch angles beyond 15° the blade-to-blade response ratios for the blade pairs with the pitched blade (Blade 3) increases significantly. This demonstrates a reduced sensitivity of this methods ability to determine damage in the presence of pitch angles beyond 15°. This observation is exemplified in Fig. 37. This figure shows the change in blade-to-blade response ratio for increased

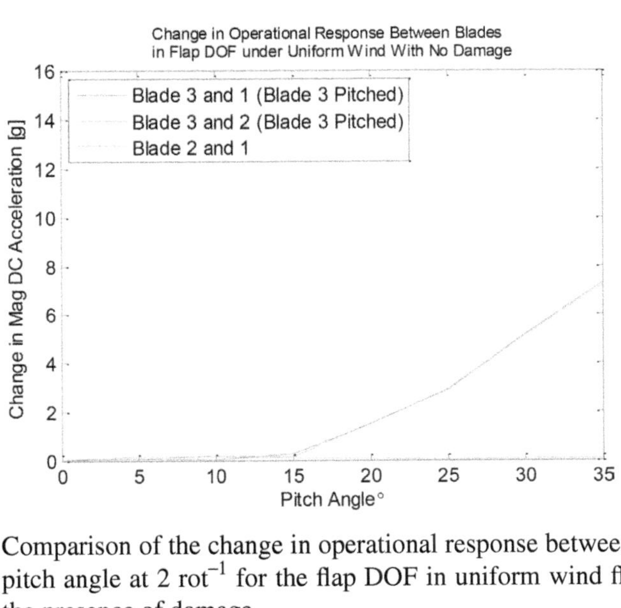

Figure 36: Comparison of the change in operational response between blades vs. pitch angle at 2 rot^{-1} for the flap DOF in uniform wind flow without the presence of damage.

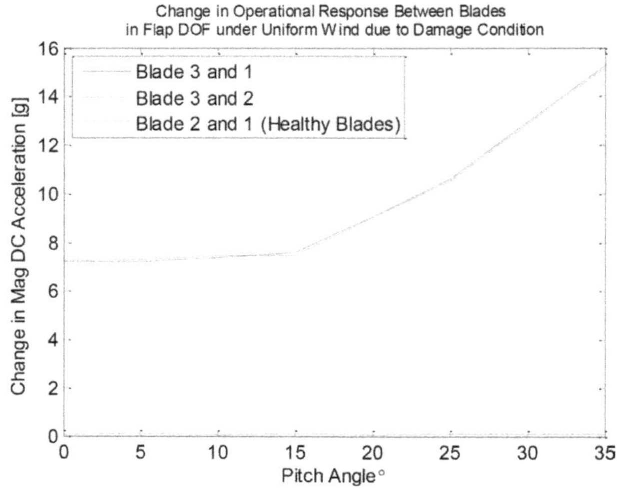

Figure 37: Comparison of the change in operational response between blades vs. pitch angle at 2 rot^{-1} for the flap DOF in horizontal shear flow when damage is present in a single blade (Blade 3).

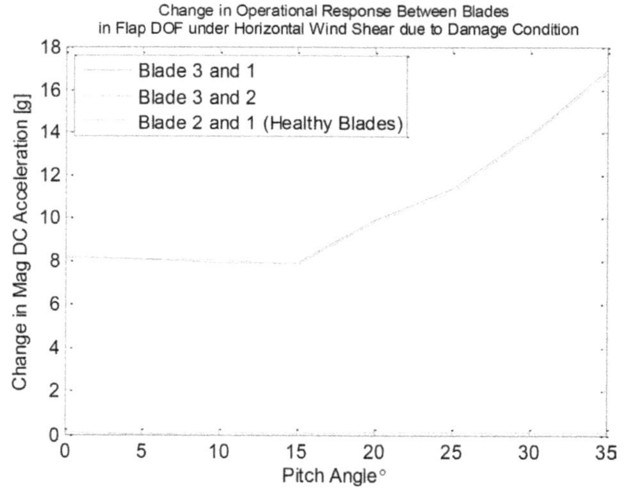

Figure 38: Comparison of the change in operational response between blades vs. pitch angle for the flap DOF in horizontal shear wind flow when damage is present in a single blade (Blade 3).

pitch error when operating in a uniform wind flow. The damaged blade (Blade 3) causes the blue and green curves to exhibit the largest changes whereas there is negligible change for the undamaged blade-to-blade response ratio (red curve). At pitch angles greater than 15° the change in the blade-to-blade response ratio is dominated by the change in pitch angle of Blade 3. Figure 38 shows a similar trend for the blade-to-blade response ratio when operating under vertical wind shear and in Fig. 39 for a turbine operating under horizontal wind shear. Again, the blue and green curves exhibit the largest changes whereas there is negligible change for the undamaged blade-to-blade response ratio (red curve).

6 Conclusions

Experiments were conducted on a small horizontal axis wind turbine by controlling the inflow conditions and measuring the inertial response of the rotor blades as a function of yaw error, pitch error and simulated damage conditions (ice accretion and blade root crack). Using cyclic averaging in conjunction with OMA techniques, it was demonstrated that the sensitivity of the measured blade flap and lead-lag acceleration responses to yaw error was larger in the presence of vertical wind shear than in the case of uniform flow, but smaller in the presence of horizontal wind shear. It was also shown using these same operational blade acceleration measurements that the presence of either horizontal or vertical shear inflow conditions makes it possible to distinguish between positive and negative yaw error, which would be important for controlling yaw error in practice. In contrast the measured blade acceleration responses did not exhibit a significant sensitivity to a single blade pitch error. Although the small-scale wind turbine blade that was tested has a substantially different aerodynamic profile than a utility-scale blade, the measurement technique and data analysis methodology are general and can translate well to utility-scale blades.

It was also shown that ice accretion could be detected using the lead-lag operational blade acceleration measurements at a frequency of one per revolution regardless of the yaw error and regardless of the inflow condition (vertical or

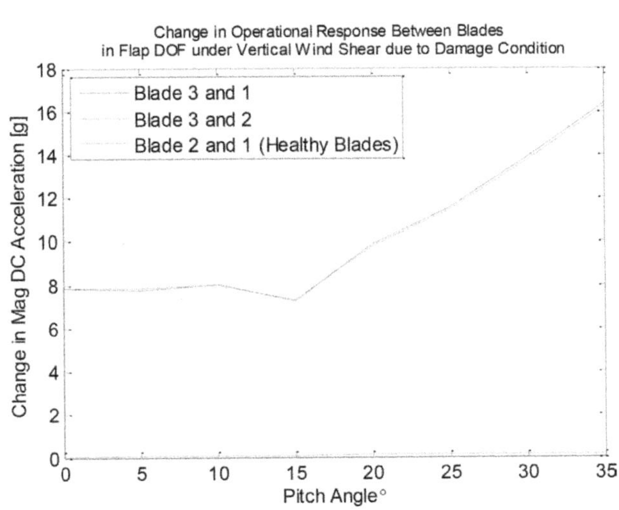

Figure 39: Comparison of the change in operational response between blades vs Pitch angle for the flap DOF in vertical shear flow when damage is present in a single blade (Blade 3).

horizontal wind shear). The percent change in the lead-lag acceleration response at this frequency was 35%. Likewise, the percent change in the lead-lag response was again 35% for pitch errors less than 5 degrees, and the presence of ice could be detected regardless of pitch error. For the case of a simulated blade root crack, the flap acceleration response of the blade showed the most sensitivity to the simulated damage. The presence of the faulty blade could be detected regardless of the yaw/pitch error and regardless of the inflow conditions.

The results of this investigation indicate that blade operational acceleration responses in conjunction with OMA techniques are a promising means of detecting and characterising yaw error and certain blade damage conditions. The experiments demonstrate that the data analysis procedure is capable of extracting features that can detect errors in yaw and pitch as well as simulated damage conditions despite significant variations in the wind inflow conditions, which is an important finding given that such variations are prevalent in utility-scale wind farms.

References

[1] Adams, D.E., White, J., Rumsey, M. & Farrar, C., Structural health monitoring of wind turbines: method and application to a HAWT. *Wind Energy*, **14(4)**, pp. 603–623, 2011.

[2] Burton, T.D., *Wind Energy Handbook*. John Wiley and Sons: New York, 2001.

[3] Robinson, M.C., Hand, M.M., Simms, D.A. & Schreck, S.J., Horizontal axis turbine aerodynamics: three dimensional, unsteady, and separated flow influences. *3rd ASME/JSME Joint Fluids Engineering Conference*, San Francisco, CA, 1999.

[4] Farret, F.A., Pfitscher, L.L. & Bemardon, D.P., Active yaw control with sensorless wind speed and direction measurements for horizontal axis wind turbines. *Proceedings of the Third IEEE International Caracas Conference on Devices, Circuits and Systems,* pp. I25/1–I25/6, IEEE, 2000.

[5] Echavarria, E., Hahn, B., van Bussel, G.W. & Tomiyama, T., Reliability of wind turbine technology through time. *Journal of Solar Energy Engineering*, **130**, pp. 0310051–0310058, 2008.

[6] Hahn, B., Durstewitz, M. & Rohrig, K., Wind energy (reliability of wind turbines experiences of 15 years with 1,500 W), eds. J. Peinke *et al.*, Springer: Berlin, 2007.

[7] Flemming, M.L. & Troels, D., New lightning qualification test procedure for large turbine blades. *International Conference Lightning and Static Electricity*, Blackpool, pp. 36.1–36.10, 2003.

[8] Ciang, C.C., Lee, J. & Bang, H., Structural health monitoring for a wind turbine system: a review of damage detection methods. *Measurement Science and Technology,* **19**, p. 12, 2008.

[9] Jureczko, M., Pawlak, M. & Mezyk, A., Optimisation of wind turbine blades. *Journal of Material Processing Technology*, **167**, pp. 463–471, 2005.

[10] Musial, W., Butterfield, S. & Mcniff, B., Improving wind turbine gearbox reliability. *2007 European Wind Energy Conference*, Milan, Italy, 2007.

[11] Mcniff, B., *Wind Turbine Gearbox Reliability*. Sandia National Labs: Albuquerque, NM, 2006.

[12] White, J., Adams, D.E., Rumsey, M. & Zayas, J., Measurement of operational loading and deflections with a smart turbine rotor blade. *Windpower*, Chicago, 2009.

[13] Seifert, H., *Technical Requirements for Rotor Blades Operating in Cold Climate*. Deutsches Windenerdie-Institut: Wilhelmshaven, 2003.

[14] Tammelin, B. & Seifert, H., Large wind turbines go into cold climate regions. *European Wind Energy Conference*, Copenhagen, 200[illegible].

[15] Hyers, R.W., McGowan, J.G., Sullivan, K.L., Manwell, J.F. & Syrett, B.C., Condition monitoring and prognosis of utility scale wind turbines. *Energy Materials*, **1(3)**, pp. 187–203, 2001.

[16] Hameed, Z., Hong, Y.S., Cho, Y.M., Ahn, S.H. & Song, C.K., Condition monitoring and fault detection of wind turbines and related algorithms: a review. *Renewable and Sustainable Energy Reviews*, **13(1)**, pp. 1–39, 2009.

[17] Beattie, A.G., Acoustic emission monitoring of a wind turbine blade during fatigue test, *Prof of AIAA*, p. 10, 1997.

[18] Zayas, J.R., Roach, D.P., Rumsey, M.A., Allan, W.R. & Horsley, D.A., (2007). Low-cost Fiber Bragg grating interrogation system for in situ assessment of structures. *Proceedings of SPIE*, **6529**, pp. 65293C.1–65293C.12.

[19] Van Hemelrijck, D., Smits, A. & Philippidis, J., Study of the behaviour of glass fibre reinforced epoxy composite system used for wind turbine rotor blades under biaxial load conditions. *Proceedings of European Wind Energy Conference Exhibition 2006*, Athens, 2006.

[20] White, J.R., *Operational Monitoring of Horizontal Axis Wind Turbines With Inertial Measurements*. Purdue University: West Lafayette, IN, 2010.

[21] Larsen, G.C., Hansen, M.H., Baumgart, A. & Carlen, I., *Modal Analysis of Wind Turbine Blades*. Riso National Laboratory: Denmark, 2002.

[22] Ghoshal, A., Dundaresan, M.J., Shulz, M.J. & Pai, P.F., Structural health monitoring techniques for wind turbine blades. *Journal of Wind Engineering and Industrial Aerodynamics*, **85**, pp. 309–324, 2000.

[23] Luczak, M., Peeters, B., Dohler, M., Mevel, L., Ostachowicz, W., Malinowski, P., *et al.*, Damage detection of wind turbine blade panels using three different SHM techniques. *Proceedings of IMAC-XXVIII,* Jacksonville, 2010.

[24] Homola, M.C., Nicklasson, P.J. & Sundsbo, P.A., Ice sensors for wind turbines. *Cold Regions Science and Technology*, **46(2)**, pp. 125–131, 2006.

[25] Barber, S. & P. Golbeck. "Wind turbine maintenance & condition monitoring." *WIND ENERGY-Technology and Planning. Bonn, Germany: The World Wind Energy Association (WWEA).* http://www. wwindea. org/technology/ ch03/en/3_4_3. html (accessed September 14 2012), 2006.

[26] Adams, D.E., *Health Monitoring of Structural Materials and Components*. John Wiley & Sons, Ltd: Chichester, UK, 2007.

[27] Shih, C.Y., Tsuei, Y.G., Allemang, R.J. & Brown, D.L., Complex mode indicator function and its applications to spatial domain parameter estimation. *Proceedings of IMAC-VII*, Las Vegas, 1989.

[28] Chauhan, S., Martell, R., Allemang, R.J. & Brown, D.L., Unified matrix polynomial approach for operational modal analysis, *IMAX-XXV*, Orlando, 2007.

[29] Rao, Singiresu S. & Fook F.Y., *Mechanical vibrations*. Vol. 4. New York: Addison-Wesley, 1995.

[30] Yutzy, J., *Characterizing the Structural Dynamic Response of a Horizontal Axis Wind Turbine in a Yawed Flow Field using Rotor and Nacelle-Mounted Inertial Sensors*. Purdue University: West Lafayette, 2011.

[31] Cattin, R., *et al.*, "Wind turbine ice throw studies in the Swiss Alps." *European Wind Energy Conference Milan*. 2007.

CHAPTER 4

Experimental Testing of Wind Turbines Using Wind Tunnels With an Emphasis on Small-Scale Wind Turbines Under Low Reynolds Numbers

Kenneth W. Van Treuren & Timothy A. Burdett
Department of Mechanical Engineering, Baylor University, Waco, TX, USA.

Abstract

The future of wind turbine design must include wind tunnel testing to produce high quality, experimental data used for design. These experimental data, both airfoil and overall system performance, can be used to validate and improve the design of the wind turbine blades and systems. Currently, little experimental testing of wind turbines is performed, with much of the aerodynamic design being accomplished using computational tools such as PROFIL and XFOIL. Computational Fluid Dynamics (CFD) prediction is improving and will be an outstanding tool for wind turbine blade design; however; these codes are not robust enough for predicting performance under low Reynolds numbers. Little experimental laboratory verification of CFD codes exists for wind turnines, especially for low-Reynolds numbers. Typically wind turbines are designed and built full scale. Because wind tunnel testing is often field tested to compare with the design predications. However, field testing can also be a very expensive process. This chapter will focus on the necessity of experimentally testing wind turbine blades to determine airfoil lift and drag data over typical Reynolds numbers of operation and also the testing of wind turbine systems (blades and generator) to determine overall wind turbine performance. This type of testing should be accomplished prior to building the full scale machine because a better design can be reached with wind tunnel testing. The Blade Element Momentum Theroy (BEMT) is typically used in the design of small-scale wind turbines and this design method depends heavily on the use of accurate airfoil data. Thus, for small-scale wind turbines, quality experimental airfoil data taken at the appropriate Reynolds numbers are necessary for accurate design and prediction of power production. The data presented are for wind tunnel testing in air; however, there is a growing interest in testing wind turbines in water tunnels, although not many facilities exist for this application. The discussions presented should be applicable to either air or water testing.

Keywords: Wind turbine experimental testing, wind turbine airfoil testing, wind tunnel, small-scale wind turbine, low Reynolds number wind tunnel testing.

1 Introduction

Having sufficient alternative energy resources is vital for the world's future. In any portfolio of future energy resources, wind energy is a logical choice to supplement existing electrical power generation. Wind energy has the potential to provide a significant portion of electrical power, however; efficiencies can be improved with research and testing of new designs. These increases in power output for current and future wind turbines will specifically occur as the result of improvements in the blade design and construction of these machines. A small body of experimental data either gathered through wind tunnel testing of airfoils or from testing full-scale wind turbines exists. Currently, many wind turbine blades are designed using 2-D airfoil data generated from numerical programs. Often experimental testing of the complete wind turbine system, both rotor and generator, does not occur except on full scale, production turbines in the field [1]. Full wind turbine systems are not usually tested in wind tunnels due to their size and the difficulties that result from both geometric scaling and Reynolds number matching. The smaller scale wind tunnels typically available at most universities do not lend themselves to testing of wind turbines. In fact, Serpa states that wind tunnels have been of limited value in wind turbine rotor development except for gathering airfoil lift and drag data [1]. Challenges in scale make testing in a small wind tunnel difficult, however; much can still be learned from testing small-scale wind turbine systems in a wind tunnel. More wind tunnel exeperimental testing of wind turbine systems needs to be accomplished.

2 Challenges of experimental testing

2.1 Design using blade element and blade element momentum theory

To understand the experimental testing necessary for wind turbines, it is first desirable to understand the design processes most widely used in wind turbine design. Blade element theory (BET) or the blade element momentum theory (BEMT or BEM) are the most commonly used aerodynamic models in wind turbine rotor design. Most wind turbine textbooks, such as *Wind Energy Explained* by Manwell, McGowan and Rogers, outline both BET and BEMT theories and give the background necessary for the development of these systems of equations [2]. These design procedures are based on the development of rotor aerodynamic theory. BET uses airfoil forces and geometry to calculate the torque generated based on free stream and rotational flows as the basis for its development while BEMT adds an actuator disk and stream-tube momentum analysis to the BET development. A representation of the stream tube used in the BEMT is shown in Fig. 1.

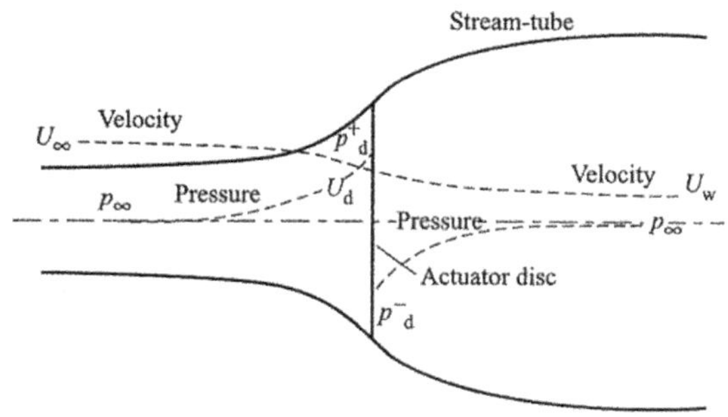

Figure 1: Actuator disc and stream-tube flow visualisation [3].

Several assumptions form the basis for the actuator disc model (based on a momentum analysis with no wake rotation). The actuator disc model assumes a homogeneous, incompressible, steady state fluid flow, no frictional drag, an infinite number of blades, uniform thrust over the disc area, a non-rotating wake and the static pressure far upstream and downstream is at the ambient static pressure. Airflow in the streamtube expands as it approaches the actuator disc resulting in a decrease in velocity and increase in static pressure. As the airflow passes through the actuator disc, energy is extracted. Immediately downstream of the disc, the static pressure decreases as a result of the energy extraction. Extraction of energy causes the flow to continue to expand with a corresponding decrease in velocity. Downstream of the actuator disc the velocity continues to decrease, which results in an increase in static pressure to ambient.

Rankine [4] in 1865 and Froude [5] in 1878 were the first to develop models for the system, that involved the simple non-rotating actuator disc as it applied to the study of ship propellers. Joukowski [6], in 1918, included the effects of a rotating disc in the analysis with wake rotation. In 1926, Betz [7] proposed his simple model to determine the power from an ideal turbine which resulted in the 'Betz Limit' used to characterise the maximum power output from the stream tube, 16/27. These momentum analyses only account for general rotor parameters and not blade geometry. Both BET and BEMT include the forces developed by the blades as a result of flow over the blades. Froude, in 1878, and Drzewiecke [8], in 1892, began the study of this development, which was refined by Glauert [9], in 1935. Glauert further refined this theory to determine the optimum design parameters to maximise power output, which was combined with the existing momentum theory [9].

In BET, the turbine blade is divided into segments of length, δr, and theoretical or experimental airfoil data are used to determine the torque and thrust on the blade. The combination of momentum theory and BET is often called BEMT or Strip Theory. The latter is named such because the momentum analysis is broken into strips that coincided with the blade segments [3]. Figure 2 illustrates both an element of a blade and a strip or annulus of the turbine rotor for that element.

Wind Energy Explained outlines the difference between BET and BEMT as simply that 'blade element theory refers to an analysis of forces at a section of the

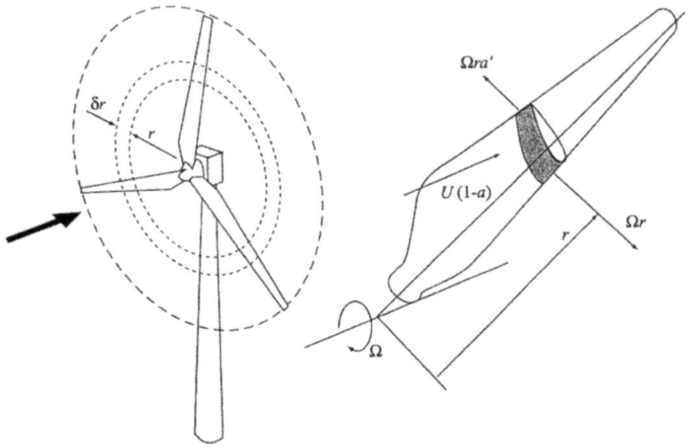

Figure 2: Blade element section and annular ring used in BET and BEMT [3].

blade, as a function of blade geometry' and 'momentum theory refers to a control volume analysis of the forces at the blade based on the conservation of linear and angular momentum' [2]. However, in some sources, such as the *Wind Energy Handbook,* this distinction is not clear and BET is often used to describe what would be considered in other sources as BEMT [3]. In *Wind Turbine Technology*, BET is described merely as a theoretical step to reach a 'modified blade element theory' [1] or what has been commonly known as BEMT.

The main assumption in BEMT is that 'individual streamtubes or strips can be analysed independently of the rest of the flow' [1]. The validity of this analysis depends on the availability of accurate 2-D airfoil data for the airfoil being used for the local Reynolds number at that particular strip. Reynolds number for this condition would be based on the relative velocity to the airfoil, which is a function of axial velocity and rotational speed for the particular strip. Analysing the strips independently of each other is reasonable if there is no significant spanwise flow. No spanwise flow would allow the use of 2-D airfoil data for each strip. However, the flow is not truly 2-D, which has led to the development of several corrections to BEMT, most notably the tip-loss analysis developed by Glauert [9]. Some additional corrections include accounting for hub-loss, stall delay in 2-D airfoil data that result from 3-D flow effects, and a skewed-wake [10].

Optimisation of wind turbine blade design using either BET or BEMT can yield different results. Each design methodology will yield a different resultant design angle of attack, α, which is directly attributable to the inclusion of the axial, a, and tangential, a′, momentum for BEMT influencing the magnitude of the relative velocity, U_{rel}. This difference can be seen in Fig. 3. Optimising the wind turbine design using BET involves selecting the angle of attack at which the lift divided by drag is maximised for the relative velocity, Urel, magnitude and flow angle, φ shown in Fig. 3a. Analysing Fig. 3b, the optimal angle of attack changes as the magnitude and direction of the relative velocity change because of the addition of the momentum terms. Since the momentum terms also change with angle of

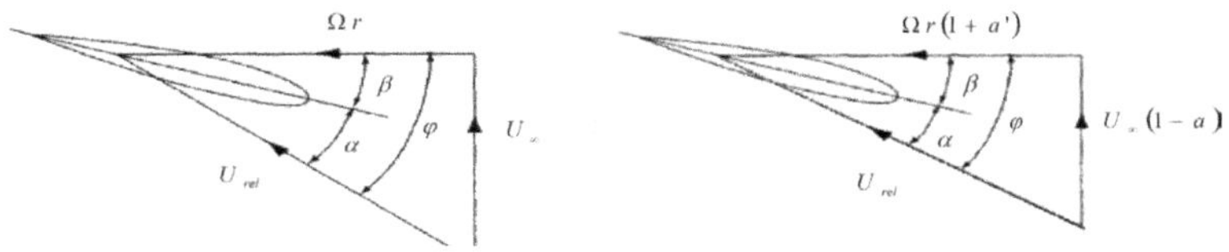

Figure 3: Elemental velocity triangle for BET (a) and BEMT (b) [3] modified by Burdett [11].

attack, the maximum power output for BEMT is not generally at the angle of attack for maximum L/D, as would be found for BET.

The two limiting factors to the accuracy of BET or BEMT are the validity of the assumptions made in the general wind turbine theory and the quality of the airfoil data used for design. BET is generally less accurate when predicting wind turbine performance. BEMT has been shown to be reasonably accurate under most conditions but depends largely on the quality airfoil data used in the design to include wind tunnel corrections applied for the data if it is experimentally obtained. Standard airfoil corrections need to be applied to airfoil wind tunnel data However, the BEMT models do not fully capture the 3D flows that occur (particularly the complex flows that occur at low Reynolds numbers). When the flow is attached to the blade, the BEMT with the quasi-standard corrections applied can be reasonably accurate. However, there is no standard set of equations that accurately model the aerodynamics when the flow is not attached, a condition which occurs at high speeds on a stall-control turbine and on wind turbine operating at low Reynolds numbers. This is particularly true for small-scale stall-controlled wind turbines, because of the sensitivity of the data collected near the stall of the airfoil in the wind tunnel. Generally speaking, however, the lack of accuracy of the airfoil data has the potential to be the largest source of error, especially for the lower Reynolds number ranges, and can mask the inaccuracies of many of the assumptions in BET/BEMT even when the proper wind tunnel corrections for the airfoil data are applied.

For higher Reynolds numbers, usually taken to be greater than 500,000, often airfoil data are generated using computational models such as the PROFIL (Eppler Code) and XFOIL. Model validation, especially over the range of Reynolds numbers that the wind turbine blade will experience, is something that is needed in the industry. This is especially true for the lower Reynolds numbers. Often actual validation for many wind turbine designs is accomplished in the field using fully operational wind turbines after they are built and installed. While this procedure is not uncommon and much can be learned from testing the actual wind turbine, it highlights the lack of appropriate experimental data to use in design and also some of the aerodynamic issues that are evident in the modelling codes [12]. Not only is wind tunnel testing of airfoils necessary to achieve an accurate design, wind tunnel testing of complete wind turbine systems (generator and blade sets) is also needed to more accurately improve the design process.

More wind tunnel testing of wind turbine systems is desirable in theory; however, there are issues with both geometric and Reynolds number scaling that must be considered. Predictions of performance from existing models can range from 25% to 175% of the measured values mostly due to the different approaches to using the 2-D airfoil data in the modelling codes [13]. This variation in performance is also dependent on the quality of the 2-D airfoil data available from either wind tunnel testing or computational design codes such as XFOIL.

While less common than theoretical analyses, several experimental studies of wind turbine systems have been performed. Blanch showed for a small HAWT, stall on the NACA 4415 blades can seem unpredictable and needs further study [12]. Sicot *et al.* examined rotation and turbulence effects on wind turbine blades emphasising stall mechanisms [14]. They showed an influence of turbulence on separation resulting in an improvement in lift and drag. More recently, Bowen *et al.* [15] tested five commercial horizontal axis small wind turbines to provide data for government agencies to consider when implementing tax incentives. They give comments on safety and function for each wind turbine system. They also report a number of data values for the turbines tested but give no insight into the aerodynamic design of the blades. More research is needed to characterise small-scale wind turbines with a focus on aerodynamic performance. NREL tested their 10-m wind turbine rotor in the NASA Ames wind tunnel and acquired data on 1,700 different test conditions [13]. These data were used to validate the predictive capabilities of various codes available to wind turbine blade designers. After conducting a 'blind comparison' of codes with their data, the conclusion was that the results were not favourable with wide variations between various code predictions. Clearly, more work is needed to adequately predict performance and further examine the aerodynamic behaviour of the blades. This implies that more experimental data are necessary to further the predictive nature of the wind turbine models and to improve the design process.

2.2 Parameters of interest in wind turbine aerodynamic testing

For either full-scale wind turbine testing or testing in a wind tunnel, there are several parameters that will be of interest to the wind turbine designer. For typical experimental procedures, a dimensional analysis would determine the dimensionless parameters of interest for the experiment. In wind turbine analysis, dimensionless parameters such as power coefficient, thrust coefficient, momentum coefficient, lift coefficient, drag coefficient and tip speed ratio would be important [2]. The power coefficient, C_P, is given in the following equation:

$$C_P = \frac{Rotor\,Power}{Power\,in\,the\,Wind} = \frac{P}{\frac{1}{2}\rho U^3 A}, \tag{1}$$

where P is typically the power from the wind turbine rotor (a function of the angular velocity Ω), ρ is the air density, U is the free stream velocity and A is the wind turbine area swept by the rotating blades. If the power, P, is considered the system power (electrical power out of the rotor/generator set), then a system power coefficient can be defined as $C_{p,s}$ [1]. The thrust coefficient, C_T, is similarly defined by eqn (2)

$$C_T = \frac{Thrust\ Force}{Dynamic\ Force} = \frac{T}{\frac{1}{2}\rho U^2 A}, \tag{2}$$

where T is the thrust force on the rotor. The momentum coefficient, C_M, is shown in eqn (3):

$$C_m = \frac{Pitching\ Moment}{Dynamic\ Moment} = \frac{M}{\frac{1}{2}\rho U^2 Ac}, \tag{3}$$

where M is the pitching moment and c is the chord of the blade. The 2-D lift and drag coefficient, C_l and C_d, are as follows in eqns (4) and (5):

$$C_l = \frac{Lift\ force_{/unit\ length}}{Dynamic\ force_{/unit\ length}} = \frac{L/l}{\frac{1}{2}\rho U^2 c}, \tag{4}$$

$$C_d = \frac{Lift\ force_{/unit\ length}}{Dynamic\ force_{/unit\ length}} = \frac{D/l}{\frac{1}{2}\rho U^2 c}, \tag{5}$$

where L is the lift, D is the drag and l is the unit length. The tip speed ratio, λ, is shown in eqn (6)

$$\lambda = \frac{\Omega R}{U}, \tag{6}$$

where R is the radius of the wind turbine. Two other parameters that are sometimes included are the advance ratio, J, and the rotor speed power coefficient, K_P, given in eqns (7) and (8)

$$J = \left(\frac{U}{v}\right)_{min} = \frac{U}{R\Omega} = 1/\lambda, \tag{7}$$

$$K_P = \frac{P}{0.5\rho R^3 \Omega^3 A} = C_P/\lambda^3, \tag{8}$$

where v is the translation (rotation) velocity. For aerodynamic testing, the Reynolds number is usually based on the aerodynamic chord of the wind turbine blade, as defined below in eqn (9)

$$Re = \frac{\rho U c}{\mu}, \tag{9}$$

where μ is the absolute viscosity.

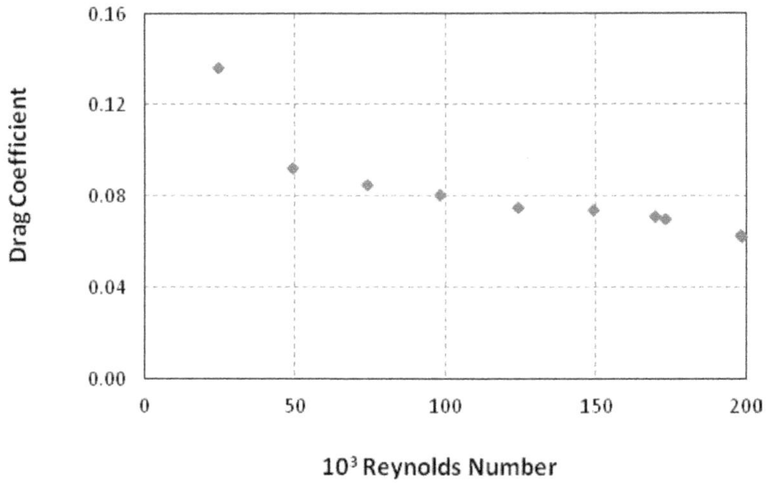

Figure 4: S818 drag coefficient vs. Reynolds number at a nominal angle of attack of 0.0° [16].

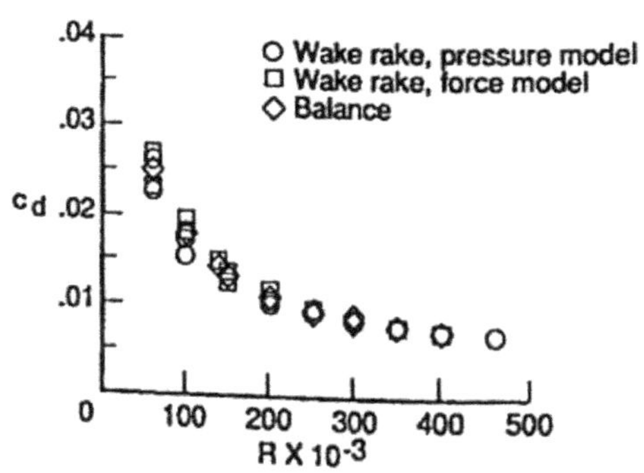

Figure 5: Drag measurements for Eppler 387 airfoil at $\alpha = 0°$ [17].

For testing of 2-D airfoils, the experimental process begins with measuring the lift and drag forces over a range of Reynolds numbers at a nominal angle of attack of 0.0° as shown in Fig. 4. This does two things. First, it defines the region where the lift and drag coefficients are reasonably insensitive to Reynolds number for the airfoil being tested as well as at what Reynolds numbers airfoil data may be an issue due to separation.

Second, especially for the lower Reynolds numbers below this region of insensitivity, measuring these values for lift and drag will be more difficult as flow separation will occur. This is illustrated in Fig. 5, which shows the difficulty of making measurements in the low Reynolds number range especially when using different experimental techniques. At higher Reynolds numbers, the onset of transition can also influence the experimental data.

For an airfoil, the coefficient of lift and drag, local Reynolds number and angle of attack are important. These values are the required parameters to quantify performance of the airfoil and acquire accurate predictions of wind turbine performance using these 2-D data. To determine these parameters, lift force, drag force, angle of attack, dynamic pressure, atmospheric temperature and atmospheric pressure must all be measured. Additionally, as part of the data analysis, appropriate wind tunnel corrections must be applied.

For particular Reynolds numbers that will be experienced over the span of the wind turbine blade, it is necessary to have the correct lift and drag coefficient data for the blade elements in the design process. Typical data for such measurements are shown in Fig. 6 for a NACA 4415 2-D airfoil.

By plotting C_l against C_d, one can determine the point of maximum L/D or (C_l/C_d) by drawing the tangent line from the origin as can be seen in Fig. 7.

Using the lift and drag data, one can also determine the point of maximum L/D by plotting L/D against angle of attack as can be seen in Fig. 8.

Thus, it is necessary to conduct an experimental airfoil test at the required Reynolds numbers over the range of angles of attack that will be experienced, including stall and beyond for stall controlled wind turbines. Local Reynolds numbers will naturally increase as the blade elements are considered from the hub to the tip of the blade for a constant chord, which might be present on small-scale wind turbines. On many larger-scale wind turbines, the chord of the blade decreases as one moves

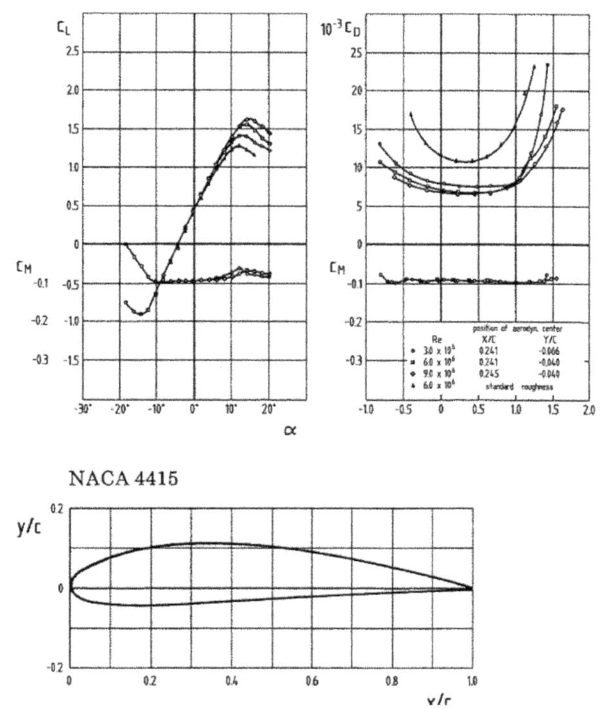

Figure 6: NACA 4415 airfoil characteristics [18].

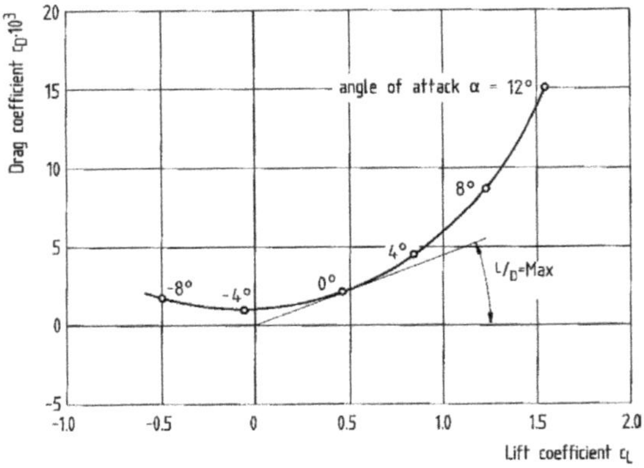

Figure 7: NACA 4415 plot C_l vs. C_d [18].

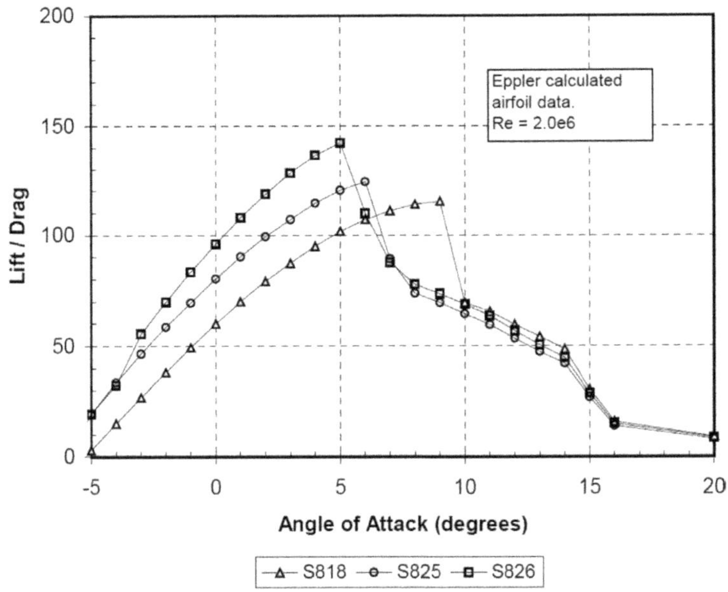

Figure 8: S818 predicted performance [19].

from the hub to the tip holding the Reynolds number constant or having it only increase slightly. The quality of the wind turbine blade design and the prediction of performance are only as good as the lift and drag data used in the design process.

3 Scaling wind turbines for wind tunnel testing

In the design of wind turbines, scaling often is a topic for discussion. Currently, wind turbine blades are designed theoretically and usually tested in full-scale turbines [15,20]. Airfoil sections are tested in the wind tunnel; however, full systems are not usually tested in wind tunnels, especially the small-scale wind tunnels found in most universities. Challenges in scale make testing in a small wind tunnel difficult, however, much can still be learned from testing full systems in a small-scale wind tunnel. Typically, a proven design of a particular size is considered the basis for scaling to a larger size for the harvesting of more power from the wind. This would be geometric scaling. Geometric scaling has been a topic largely ignored resulting in a relatively little literature on the subject, especially as it applies to wind turbines [21]. The topic of scaling is briefly covered in most wind energy textbooks; however, scaling relationships are considered mainly for changing the geometric size of the turbine. Reynolds number matching is ignored [22]. For scaling relations, Manwell *et al.* [2] use the following assumptions:

1. Keep the tip speed ratio constant.
2. Keep the blade profile, the number of blades and the blade material constant.
3. Geometric similarity is maintained to the extent possible.

Table 1 summarises the relationships between quantities of interest for wind turbines based on rotor diameter as the scaling factor. Using these guidelines if the rotational speed is halved then the radius is doubled, keeping the tip speed ratio constant. If the tip speed ratio is constant, then the power, thrust and momentum coefficient will remain constant as these are the functions of tip speed. A constant tip speed also preserves the angle of attack for the wind turbine blade resulting in the same lift and drag coefficient. The danger in using the concept of geometric

Table 1: Wind turbine scaling laws without Reynolds number matching [2].

Quantity	Symbol	Relation	Scale dependence
Power, forces, and moments			
Power	P	$P_1/P_2 = (R_1/R_2)^2$	$\sim R^2$
Torque	Q	$Q_1/Q_2 = (R_1/R_2)^3$	$\sim R^3$
Thrust	T	$T_1/T_2 = (R_1/R_2)^2$	$\sim R^2$
Rotational speed	Ω	$\Omega_1/\Omega_2 = (R_1/R_2)^1$	$\sim R^{-1}$
Weight	W	$W_1/W_2 = (R_1/R_2)^3$	$\sim R^3$
Aerodynamic moments	M_A	$M_{A,1}/M_{A,2} = (R_1/R_2)^3$	$\sim R^3$
Centrifugal forces	F_c	$F_{c,1}/F_{c,2} = (R_1/R_2)^2$	$\sim R^2$
Stresses			
Gravitational	σ_g	$\sigma_{g,1}/\sigma_{g,2} = (R_1/R_2)^1$	$\sim R^1$
Aerodynamic	σ_A	$\sigma_{A,1}/\sigma_{A,2} = (R_1/R_2)^0 = 1$	$\sim R^0$
Centrifugal	σ_c	$\sigma_{c,1}/\sigma_{c,2} = (R_1/R_2)^0 = 1$	$\sim R^0$
Resonances			
Natural frequency	ω	$\omega_{n,1}/\omega_{n,2} = (R_1/R_2)^1$	$\sim R^{-1}$
Excitation	Ω/ω	$(\Omega_1/\omega_{n,1})/(\Omega_2/\omega_{n,2}) = (R_1/R_2)^0 = 1$	$\sim R^0$

similarity is that while the volume of a wind turbine will increase as the scaling factor cubed, the power production is a factor of the scaling factor squared.

These relationships are valid providing the weight does not become the prime concern for wind turbine blades (as might be true for large turbines) and that the values of design parameters should not result in a critical Reynolds number of less than 200,000 [18]. Reynolds numbers less than 200,000 result in testing problems for small wind turbines (D < 5 m) or wind tunnel testing [23]. These relationships are also only true if the blades are scaled up or down exactly in proportion to the change in radius. Peterson states that small variations in the radius, such as extensions of the blade supports at the hub, cause the scaling projections to be inaccurate [24]. Items associated with wind turbines that are considered in scaling are the rotor, weights in the nacelle and the tower. Jackson has shown that fatigue loads are able to scale but only if dynamic and aerodynamic similarity is maintained [20]. Shrestha *et al.* have even determined scaling laws to estimate weights for direct drive generators of different power levels used with wind turbines [25]. Associated with the scaling would be the increased weight of the blades and the loads generated on the blades themselves. This does not always follow the same geometric similarity. Moe states, for example, that increasing the size by 20% would increase the power production by 50% while increasing the weight by 73% [22]. Increasing the size of the wind turbine using geometric scaling will usually find that cost do not scale in the same proportion and will become an increasingly larger part of the overall energy production costs [22]. It is suggested that scaling a wind turbine up past a certain size will eventually not be possible and will result in a constant response level, regardless of size [22]. Manwell *et al.* [2] state in their third scaling relation that 'geometric similarity is maintained to the extent possible' recognising there may be cases where geometric similarity is not possible.

Van Treuren and Gregg discuss the importance of dimensionless coefficients for scaling, such as power coefficient, thrust coefficient, momentum coefficient, lift coefficient, drag coefficient and tip speed ratio [26]. The other non-dimensional number of importance when considering similarity or scaling is the Reynolds number based on the chord for the wind turbine blade. Burdett and Van Treuren have looked at the case for scaling when Reynolds number matching is included [27]. Table 2 shows that values, such as the rotational velocity, scale as one over

Table 2: Wind turbine scaling laws (scaling in same fluid) [27].

Parameter	Symbol	Relation
Radius	R	$R_1/R_2 = sc$
Chord	C	$C_1/C_2 = sc$
Kinematic viscosity	ν	$\nu_1/\nu_2 = sc$
Freestream velocity	U	$U_1/U_2 = 1/sc$
Rotational velocity	Ω	$\Omega_1/\Omega_2 = 1/sc^2$
Rotor power	P_{Rotor}	$P_{\text{Rotor1}}/P_{\text{Rotor2}} = 1/sc$
Torque	τ	$\tau_1/\tau_2 = sc$

the radius squared instead of one over the radius as seen in Table 1 for rotational speed. As can be seen in Table 2, the freestream velocity must increase by the geometric scaling factor, and the rotational velocity must increase by the square of the scaling factor. Because of the relationships derived from Reynolds number matching, scaling in this manner can lead to impractical freestream and rotational velocities.

While matching Reynolds numbers should also occur in experimental testing, matching Reynolds numbers leads to unrealistically high free stream and tip velocities when scaling from a large turbine blade to a smaller scaled wind turbine blade for testing in a wind tunnel. The Reynolds number is also not a feasible way to scale wind tunnel turbines to a large scale. At higher Reynolds numbers, above 500,000, the aerodynamics do not change significantly with Reynolds numbers. To scale down with the Reynolds number, the wind tunnel speed would have to be unrealistically high, making testing nearly impossible. To scale up would result in low speeds and low Reynolds numbers. For example, the wind tunnel testing of a 0.508 m (20 in.) diameter fan at 64.37 kph (40 mph) when scaled to a 2.4383 m (8 foot) diameter turbine blade would have a scaled wind speed of 13.41 kph (8.33 mph). Since many of these smaller wind turbines with 2.4383 m (8 foot) diameter turbine blades are optimised for approximately 40.23 kph (25 mph), the corresponding speed for reducing the size of the turbine to that needed for testing in the wind tunnel would be unrealistically high.

The result is that to test the full-scale Reynolds number on a smaller scale, it is necessary to test the at tip speeds above the full-scale value. Wainauski tested a 1/30 scale model of a 3.5-MW wind turbine and did match Reynolds numbers. The result was test tip speeds of 183 m/s (600 ft/s). This led to the recommendation by Wainauski to keep the tip speeds below the drag rise Mach numbers of the blade airfoils to avoid compressibility losses [28]. Gao and Hu recognise that dynamic similarity (Reynolds number matching) is hard to achieve with wind turbines of different sizes. They show that the power coefficient of two wind turbines is different if the Reynolds numbers are different. If the sizes of wind turbines are different, such as is usually the case with testing a scaled wind turbine in a wind tunnel as opposed to an actual wind turbine, the Reynolds number cannot be satisfied and wind tunnel data cannot be directly applied to the original wind turbine [29]. An important impact of Reynolds number is its effect on the 2-D airfoil data used in the design of the wind turbine. Gao and Hu have found that the lift and drag coefficients changed little in the Reynolds number range of 200,000–2,500,000 [29]. Burdett *et al.* have also shown an effect of Reynolds number on design considerations for wind turbines using 2-D air foil data for lower Reynolds numbers [30]. They show that at lower Reynolds numbers, the angle of attack for maximum C_L/C_D changes and this must be considered when using BET. Even more important is the onset of flow separation at Reynolds numbers below 100,000, which must be addressed by active or passive flow control devices if a reasonable power production with the wind turbine is desired.

Design procedures are important for the design of any wind turbine. Wind turbines of all sizes are available for consumer use. Large turbines with megawatt capacity can have wind turbine blade diameters from 80 to 100 m while small, residential wind turbines of 1–2 kW can have blades with rotor diameters of 3 m or less [23]. However, it is extremely important to have design procedures to take advantage of information gathered for smaller wind turbines under laboratory conditions that can be scaled to the larger diameters. It is also advantageous to test scaled-down models in a wind tunnel to examine parameters such as torque, power and thrust characteristics planned for larger systems. This experimental testing would help to minimise risks to the maximum extent possible.

While often difficult, wind tunnel testing of wind turbines can provide very meaningful results and minimise research and development risks. For example, wind tunnel data can be used for optimising designs. Gregg *et al.* studied the effect of tip speed ratio (Ω, defined in Eq. (8)), number of blades and surface roughness on wind turbine performance in a wind tunnel with a 2 ft (0.61 m) square cross-section [16]. Additionally, wind tunnel tests can be used to develop and validate aerodynamic modelling tools such as the Unsteady Aerodynamics Experiment (UAE) and the Model Experiments in Controlled Conditions (MEXICO) wind tunnel tests [31]. Other tests of this nature have been performed by Ronsten [32] and Snel *et al.* [33] to better model the stall delay phenomena.

In wind tunnel testing of wind turbines, scaling is an integral topic for discussion. For wind tunnel testing, a turbine design is typically scaled down to allow testing in a wind tunnel without significant blockage. As noted by Simms *et al.* in the case of the UAE and also the MEXICO tests, however, it is sometimes necessary to test a full-scale turbine because scaling is impractical [13]. Nonetheless, full-scale testing in a wind tunnel is difficult because typical wind tunnel sizes are too small to fit most wind turbines while keeping blockage minimal.

When scaling a turbine for design or wind tunnel testing, typically only geometric scaling and TSR matching are applied. Multiple studies have been performed using only these scaling criteria. For example, Cho *et al.* tested a 12% scaled model of the NREL Phase VI turbine and analysed the acoustic noise produced [34]. Chen and Liou tested multiple scaled models and analysed the significance of the blockage correction with wind tunnel blockage between 10.2% and 28.3% [35]. Jackson studied the scaling of fatigue loads with good correlation for most of the conditions studied [20]. Hulskamp *et al.* scaled trailing edge flaps to determine their ability to reduce dynamic loading [36].

Another important aspect of scaling is Reynolds number matching (Reynolds number based on chord as defined in eqn (9)). Without Reynolds number matching, the aerodynamic conditions between the two turbines are not truly identical. However, for Reynolds numbers greater than 500,000, airfoil performance does not change significantly with Reynolds number so Reynolds number matching is less significant. Most medium to large turbines (>20 m) operate at higher Reynolds numbers, and this effect does not need to be taken into account [1]. This is demonstrated by Jackson, who documents that fatigue loads for larger turbines can be scaled without Reynolds number matching and reasonable agreement can be achieved [34]. At Reynolds numbers below 500,000, however, 'extreme and unusual behavior is caused by anomalous transition, separation and bubble formation phenomena' [1].

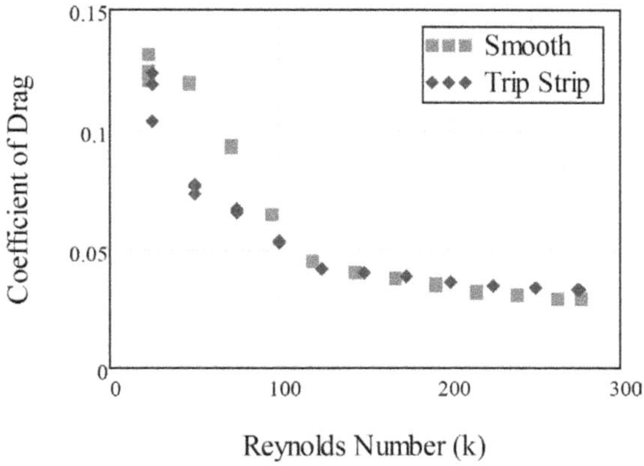

Figure 9: S823 drag dependence on Reynolds number [27].

Because of the unique aerodynamics at Reynolds numbers below 500,000, the flow can vary significantly with both flow and geometric parameters [17]. Burdett *et al.* discuss how airfoil lift and drag coefficients change with Reynolds number and the effect that change has on wind turbine design. In particular, they show that the optimal angle of attack for wind turbine design varies with Reynolds number [30]. Additionally, the flow is very sensitive to changes in freestream turbulence, surface finish and airfoil shape [37]. This sensitivity to roughness can be exploited for low Reynolds numbers to reduce the separation that occurs and likewise increases both airfoil and wind turbine performance [17,38]. Figure 9 shows how drag coefficient for the S823 airfoil can be significantly affected by roughness. In this test performed at approximately zero angle of attack (α), the roughness applied is a trip strip with a height of 0.35% of the chord length and a width of 2.0% of chord length, located at 25% of the chord on the upper airfoil surface only. At Reynolds numbers below 125,000, the addition of the trip strip significantly reduces drag up to 36.9%. However, above this Reynolds number the drag is actually increased by up to 13% (at 275,000).

Due to the sensitivity of this flow regime, the specific Reynolds number below which the unusual flow phenomena occurs is also sensitive to similar changes. While this Reynolds number is generally taken to be 500,000 [1], Gao and Hu predicted in a computational fluid dynamics (CFD) model that this number should be closer to 200,000 and that the airfoil lift and drag coefficient changed little in the Reynolds number range of 200,000 to 2,500,000 (specific to the rotor and conditions modelled) [29].

Two situations exists when scaling down a turbine for wind tunnel testing that will result in Reynolds numbers below 500,000 and necessitate Reynolds number matching. First, if the original full-scale turbine already experiences low Reynolds numbers, Reynolds matching must be used in scaling, as in the testing of a scaled 1.15-m turbine optimised for 5 m/s by Burdett *et al.* [39]. Second, if scaling down the turbine results in Reynolds numbers below 500,000, either Reynolds number matching must be used or some other test parameters must be controlled such that the Reynolds numbers on the blade remain above 500,000. One such test was performed by Wainauski in scaling a 3.5 MW turbine by a 1:30 ratio to an 8 ft (2.44 m) diameter for testing in 18 ft (5.49 m) octagonal test section. In this test, a TSR was chosen to keep the Reynolds numbers over the blade near 600,000 and the tip speed below 183 m/s. Because of this, Reynolds number matching was not necessary and compressibility effects did not need to be taken into account. However, this study was very beneficial in demonstrating the effects of scaling, surface roughness and blade shape modifications on the full-scale 3.5 MW turbine. After correcting the data for scaling effects, the wind tunnel data agreed with the full-scale annual energy production (AEP) within 1% [28].

A few other considerations exist for scaling. An obvious parameter for scaling is mass, which is difficult to scale. For steady-state wind turbine tests in smaller wind tunnels, however, not scaling the mass will not change the measured values. Additionally, the starting torque for the full-scale generator is very difficult to scale. For this reason, cut-in speed (the wind speed at which the turbine begins to spin) cannot be scaled very easily. Peterson notes that small variations in turbine geometry, such as radius, may cause the scaling projections to be inaccurate [24].

4 Wind tunnel testing

The use of wind tunnels in testing wind turbines is highly advantageous. However, wind turbine testing of actual wind turbine systems with large blades in a wind tunnel is not usually possible because the large diameters do not physically fit into a wind tunnel or the rotor disc will cause an unacceptably large blockage in the tunnel. It is possible to test fixed, full-scale wind turbine systems at locations such as the National Renewable Energy Laboratory's National Wind Technology Center in Colorado but this is costly [40]. Other testing of wind turbine systems outside a wind tunnel has been reported by Blanch [12]. Blanch tested a 3-m diameter wind turbine mounted to a trailer by towing the trailer along an airport runway. He noticed some variability in the data due to differing environmental conditions. In a wind tunnel, both the wind speed and quality are carefully controlled, eliminating the variability. The largest tunnel in the literature used in wind turbine testing is the NASA Ames Research Center's 34.4 m (80 ft) × 36.6 m (120 ft) wind tunnel [13]. This tunnel enabled testing of a 10-m diameter wind turbine system (NREL Phase VI test) keeping the blockage to an acceptable level so that corrections were not necessary (<10%). Some large wind tunnels do have dimensions of 10 m × 10 m but even this is not large enough for most of the wind turbine systems [18]. The majority of wind tunnels are much smaller and, if the blockage constraint is satisfied, this limits the size of the turbine that can be tested. The MEXICO tests were done in the DNW facility with a 4.5-m diameter turbine in a 9.5 m × 9.5 m wind tunnel test section with no corrections [31]. Sicot *et al.* tested a 1.3-m diameter turbine in a 4 m × 4 m test section with no blockage corrections [14]. Small wind tunnels of the sort found in research universities are also useful for wind turbine system testing. Burdett *et al.* have successfully tested systems in a tunnel with dimensions of 0.6 m × 0.6 m with wind turbine diameters of 0.5 m [27,39]. For these tests, blockages corrections must be applied. Wind tunnels can still be of use to examine specific aspects of wind turbine blades to develop technologies to apply to the wind turbine blades. Usually, wind turbine tests involve the measurement of power out and how a rotor would respond to unsteady conditions [18]. For power measurements under steady conditions, the model does not have to match the material properties of the actual turbine blade; however, to match dynamic response to unsteady conditions, the material elastic properties should be similar.

Wind turbine testing involves wind tunnels in a variety of ways. Maeda *et al.* [41] used an actual full-sized wind turbine blade to measure the static pressure distribution over the blade. This was a 10-m diameter wind turbine rotor with the hub located 13 m above the ground. This pressure distribution was then compared with a similar airfoil in a Gottingen-type wind tunnel. Comparisons were made with good results. Butterfield *et al.* [42] made a similar test of airfoil cross-sections in a large-scale wind tunnel, comparing with other wind tunnel and actual C_L data with good results. Currently, new wind turbine designs are typically tested in the field by measuring the power output and developing a power curve for the actual turbine. It is difficult to measure the aerodynamic characteristics of the wind turbine in the field.

Some testing has been done on small-scale models in wind tunnels of 1 m × 1 m or less [27, 35, 43–48]. Blockage is always a consideration under these small-scale conditions. Traditionally, tunnel blockages of 6%–7.5% or even up to 10% have been acceptable according to Howell *et al.* [46]. Scaling down rotating machinery typically leads to losses in efficiency which must be understood [49].

Evident from the small-scale models in wind tunnel experiments is the low-power output from the wind turbine systems. This is due to a number of factors. First, tip speed ratios are lower due to the smaller diameters. Attempting to match Reynolds numbers is not possible and would lead to unrealistic rotational speeds. Reynolds numbers based on the chord leads to numbers that are well below the 200,000 previously mentioned as a lower limit. Second, the smaller size of the turbine means there is less power output or power available. The result is low C_P values for the tip speed ratios tested. Third, the smaller chord for the wind turbine blade also reduces the amount of lift generated by the blade and thus, the power output is lower. Fourth, finding an appropriate small-scale generator for use in the wind tunnel is difficult. Several wind tunnel experiments were presented in the literature without significant discussion of the choice of generator. Fifth, designing a fixed pitch turbine blade leads to an optimised performance for only one speed. Operating above or below this speed leads to lower power generation. Lastly, there are mechanical losses of the system, which are hard to evaluate. Often this is evident in the wind tunnel by the higher free stream airspeed necessary to start the wind turbine rotating for testing. In spite of these challenges, there is a place for measurement of full wind turbine systems in a small-scale wind tunnel.

4.1 Testing airfoils

4.1.1 Airfoil characteristics

Much of the literature concerning wind tunnel testing of wind turbine systems involves experiments to determine the lift and drag coefficients of airfoil cross-sections that will be used in wind turbine blade design. Lift and drag data for airfoils are generally found either by experimental testing or by numerical simulation. Experimental testing by NACA in the 1930s resulted in the family of airfoils used in the design of aircraft [50,51]. Early wind turbines began using these NACA airfoil sections such as the NACA 4415 airfoil [18]. These airfoils were considered appropriate because of their laminar flow characteristics. For wind turbines, it is desired to maximise lift and minimise drag for the chosen design speed. Thus, for BET design, operating at maximum L/D ratio (minimum drag for the lift produced) is desirable. Figure 6 shows data for the NACA 4415 airfoil, typical for these types of tests. These airfoils are appropriate for most conditions because of their low L/D. However, the surface finish of the blades must be very smooth as roughness can cause a laminar boundary layer to prematurely transition to turbulent flow. Additionally, an increase in the freestream turbulence can have a similar effect. For this reason, differences in surface roughness of airfoil models and wind tunnel freestream turbulence between experiments can cause differences in data. Additionally, the blockage corrections applied and the measurement methods can affect the data [12]. From the plot of C_L vs. C_D shown in Figs. 7 and 8, it is possible to determine the maximum L/D ratio directly.

Wind tunnel testing of airfoil cross-sections will always have its place in the design of wind turbines. While NACA airfoil cross-sections are still being used on wind turbines, attention is being given to designing airfoils that are more appropriate for wind turbines. Figure 10 shows data for a laminar flow LS-1 airfoil. Because the local Reynolds number over the blade is a function of the blade radius, data must be collected for the entire range of Reynolds numbers expected. The National Renewable Energy Laboratory (NREL) has supported research in this area of airfoil design [52]. Other researchers are developing alternative families of airfoils and testing NREL airfoils for wind turbine use [19,53–56]. Future work with wind turbine airfoils will include modification of existing airfoils for better performance [57,58].

Another area which is critical for wind turbine performance is flow separation, which can occur under low Reynolds number conditions below 500,000. Modern computational design programs are questionable in the low Reynolds number regime and tend to under predict the laminar separation bubble drag [53]. Also, the effects of leading edge roughness is another area that needs further research.

Several studies have been accomplished to gather experimental data for low Reynolds numbers and compare to simulations. McGhee *et al.* studied the Eppler 387 airfoil at Reynolds numbers from 60,000 to 460,000 and presented both experimental and simulation data using PROFIL, a numerical design code developed by Prof. Richard Eppler [60]. A good comparison was found between the theoretical and experimental results for the higher Reynolds numbers, but at lower Reynolds numbers PROFIL was unable to accurately predict the increase

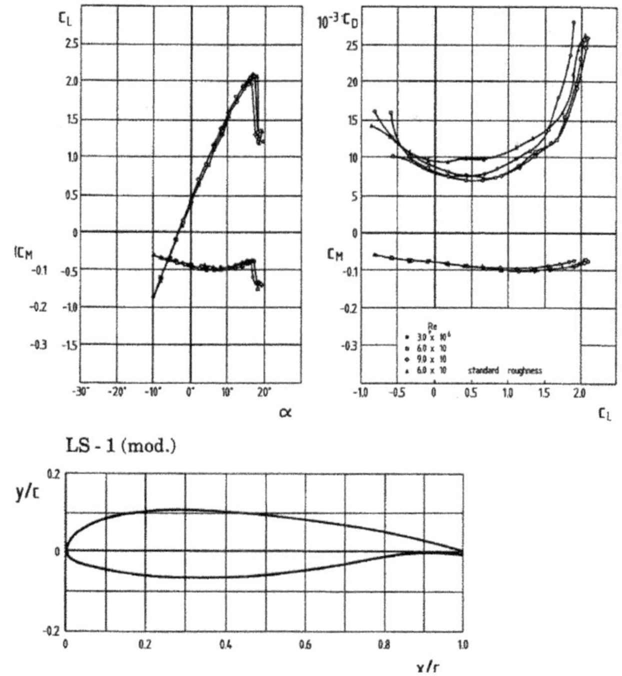

Figure 10: LS-1 laminar flow wind turbine airfoil [59].

in drag resulting from a laminar separation bubble. They concluded that the simulation would predict a lower drag coefficient than was measured experimentally. The laminar separation on low Reynolds numbers airfoils was studied by Hu and Yang [61]. They document the behaviour of this separation bubble for a Reynolds numberof 70,000 using both PIV and surface pressure measurements. It was found that the separation bubble moves forward as the angle of attack is increased. Being able to predict this separation bubble and its movement is critical to predicting airfoil performance for small-scale wind turbines. A number of studies have been done on airfoils for small-scale wind application [62–69]. A wide range of Reynolds numbers have been tested and/or simulated with these studies. For Reynolds numbers above 200,000, agreement between experimental data and PROFIL was generally good. However, at 100,000 and below, where laminar separation can dominate the flow over an airfoil, the agreement was considered fair [64]. Selig and McGranahan [69] have shown that drag behaviour is highly dependent on Reynolds number, especially in low Reynolds number regimes. Thus, there is a need for additional airfoil analysis, both experimental and simulated, at the low operational Reynolds numbers for small-scale wind turbines to further refine blade design.

Two notable sources of airfoils designed for the low Reynolds numbers that small-scale wind turbines experience are the National Renewable Energy Laboratory (NREL) as well as Dr. Michael Selig at the University of Illinois, Urbana-Champaign

(UIUC). NREL has developed two families of airfoils (each family is for a specific diameter turbine) for turbines under 10 m in diameter. The S833–S835 have been designed for 1–3 m turbines [70], and the S822–S823 have been designed for 3–10 m turbines [71]. These airfoils have been designed theoretically using PROFIL in the Reynolds numbers range of 150,000–800,000. Among other design criteria, these airfoils were designed to minimise the effect of leading-edge roughness by simulating the flow transition that occurs under those conditions.

Additional airfoils have been designed by Selig *et al.*, and these as well as other low Reynolds numbers airfoils have been tested experimentally over Reynolds numbers range of 30,000–500,00 [72–75]. In sources [73] and [74], the same airfoils were tested with both normal flow conditions and tripped flow conditions (to simulate leading edge roughness). For most airfoils testing, however, data were only taken at a Reynolds number as low as 100,000. Additionally, the data were never taken post stall, which is necessary for the design of stall-regulated wind turbines.

If lower Reynolds number or post-stall data were desired, experimental testing must be performed because neither PROFIL nor XFOIL can accurately predict these conditions on all airfoils due to flow separation over most of the airfoil surface under these conditions. While CFD could be an alternative to experimental testing, it must first be validated experimentally before the theoretical model can be extrapolated to conditions it cannot currently predict. CFD does not presently model transition from laminar to turbulent well, which is a critical phenomenon in a separation bubble as it causes reattachment. Since an accurate model does not exist, experimental testing is still required.

4.1.2 Airfoil testing in a wind tunnel

While airfoil testing has become relatively standardised, many differences exist in wind tunnel sizes that have been used to measure airfoil data. Large tunnels, such as the tunnel at the University of Illinois Urbana Champlain (UIUC), have test sections as large as 3 ft × 4 ft. These types of tunnels employ several different techniques to measure the forces on the airfoil. A common method used is the force balance, or sting, mount which measures lift and drag directly. Some tunnels do measure lift and drag using a side mount system which essentially isolates the flow to be 2-D. Other experiments measure drag downstream of the model by characterising the wake and using the momentum deficit approach. What is evident is that experimental data gathered are highly dependent on the wind tunnel and the method used to measure the forces. Experiments do not usually report the turbulence intensity of the flow, which also can have a large impact on the separation characteristics of the airfoil.

A typical small laboratory wind tunnel, such as seen in Fig. 11, is an open-circuit tunnel with a 24-in × 24-in × 48-in test section. The Model 406 Wind Tunnel was manufactured by Engineering Laboratory Design, Inc. and has a 40-hp electric motor that drives a constant pitch fan. Varying the RPM gives a low velocity of 0.1 m/s to an upper tunnel velocity greater than 50 m/s. Tunnel velocity variation over the test section is less than ±1%. The contraction ratio from the inlet

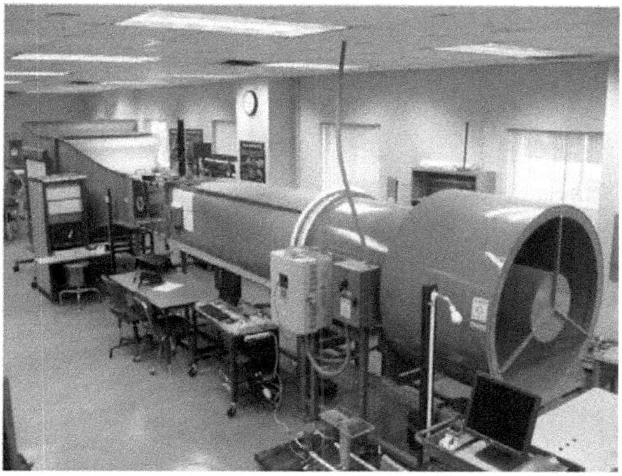

Figure 11: Typical small-scale wind tunnel [27].

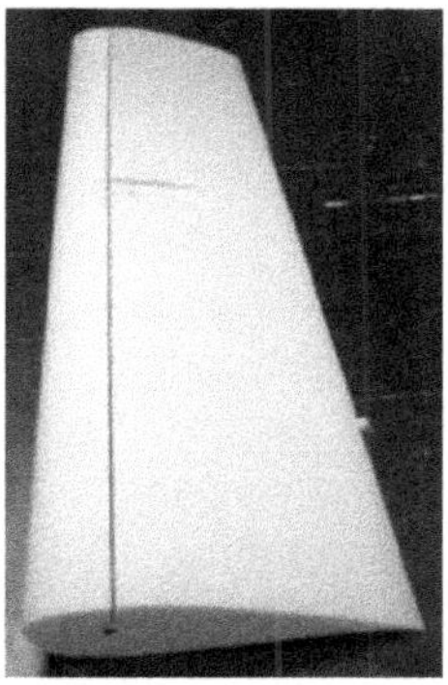

Figure 12: ACP Spyderfoam Airfoil [76].

is 6.25:1. The contraction ratio, a precision honeycomb inlet and three graduated, high-porosity screens provide a clean inlet turbulence intensity of approximately 0.2%. Access to the tunnel test section is provided through a removable floor and ceiling.

An S823 airfoil is shown in Fig. 12 and is typical of airfoils tested in this wind tunnel. The test article is 24 in. in span with a 6-in chord, and is manufactured out of ACP SpyderFoam. A hole was cut through the centre of the airfoil, which spans the airfoil at the quarter-chord for a 3/16-in. stiffener rod. The groove on the top of the airfoil is remaining from cutting the hole for the stiffener rod and was shown not to affect the airfoil performance. The airfoil also has a groove cut chordwise on the airfoil centreline on both the top and bottom surfaces to allow the mounting system to be flush with the airfoil (Fig. 13) and mounts directly onto the Engineering Laboratory Design (ELD) force balance (Fig. 14). The airfoil spans the test section to minimise and 3-D flow effects.

Figure 13: Mounting system [76].

The lift and drag were measured using the ELD dynamometer. Its DC voltage output was connected to a National Instruments BNC 2110 and recorded using Labview. The dynamic pressure was measured using a pitot-static tube upstream of the airfoil. The angle of attack was measured using a digital torpedo level, accurate to 0.1°.

4.2 Testing wind turbine systems

Full wind turbine systems tests can occur in many size wind tunnels. The purpose of each of these tests is to gather experimental data under reasonable conditions that can be compared with CFD or be scaled up (or down) to appropriate size turbines. Small-scale wind turbine tests using wind tunnels of the size depicted in Figs. 11 and 14 are of value in determining wind turbine system characteristics. Burdett *et al.* have successfully used this is size wind tunnel to compare BET and BEMT design methods as well as geometric and Reynolds number scaling [39,27]. Other tests can be conducted on tip speed ratios, number of blades, optimisation of twist, effects of roughness, etc. such as was studied by Gregg and Van Treuren and Gregg et al. [77,16]. Typical setup for the wind turbine system in the wind tunnel is shown in Fig. 15. Blades for these tests were designed in Solidworks and printed on an OBJET 30 3-D printer and can be seen in Fig. 16.

The results of these experimental tests provide valuable information. Figs. 17–19 show the results of some these testing. The graphs illustrate a comparison between tested parameters, where $C_{P,S}$ is the wind turbine system electrical power output normalised by the power available in the air. From these tests, it is easy to compare the effects of tip speed ratio, number of blades and the addition of roughness to the blades on power output for a wind turbine rotor/generator combination.

Figure 20 shows the comparison of the rotor power coefficient of two-rotor blade sets designed using either BET or BEMT (seen in Fig. 16b). Figure 21 shows the rotor power coefficient when applying geometric and Reynolds number scaling of the wind turbine blades shown in Fig. 16a. Obviously, wind tunnel testing of full-wind turbine systems provide insight into the design of wind turbine rotors.

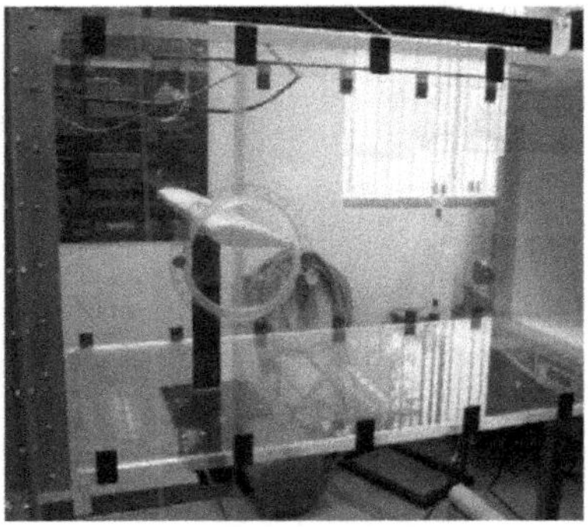

Figure 14: Airfoil and test section with force balance [76].

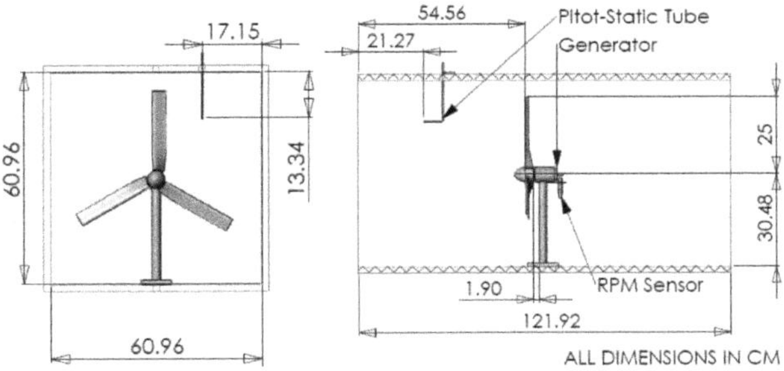

Figure 15: Dimensions of turbine and wind tunnel test section [39]

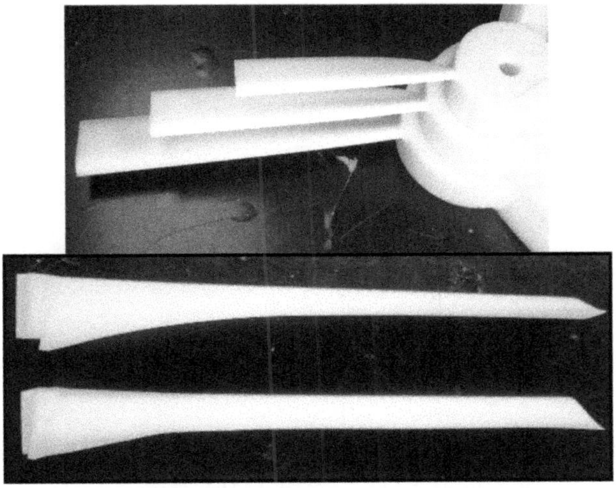

Figure 16: Printed wind turbine blades used in scaling tests with dovetail hub and showing different angles of twist for BET and BEMT blades [27,39].

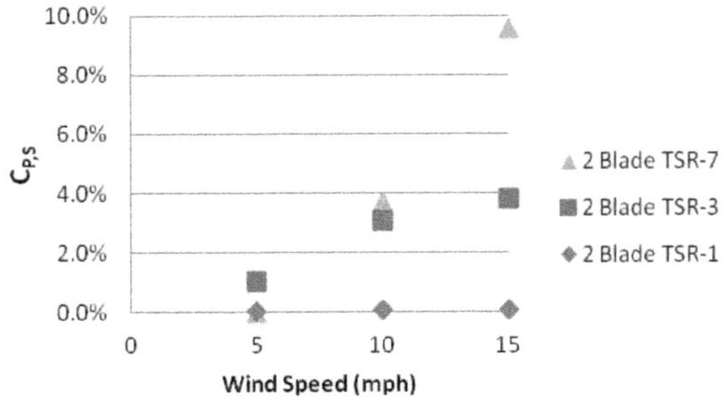

Figure 17: S818 two blade tests [16].

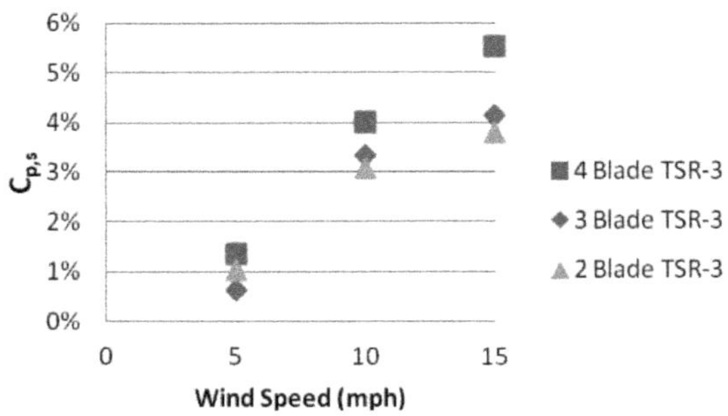

Figure 18: S818 performance for different blade sets at a constant TSR of 3 [16].

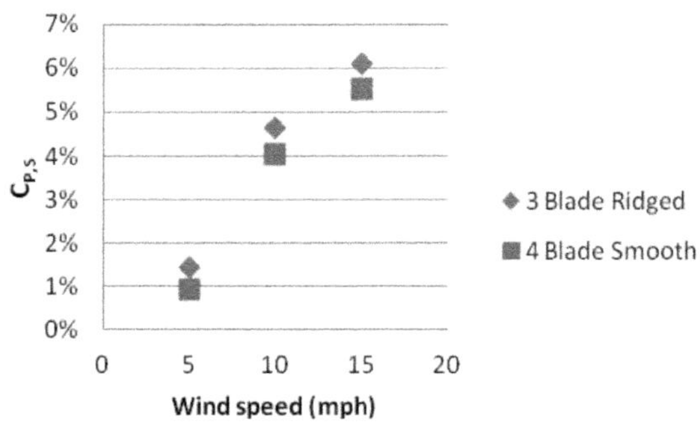

Figure 19: S818 data comparing a 4-blade smooth configuration with a 3-blade ridged configuration at a TSR of 3 [16].

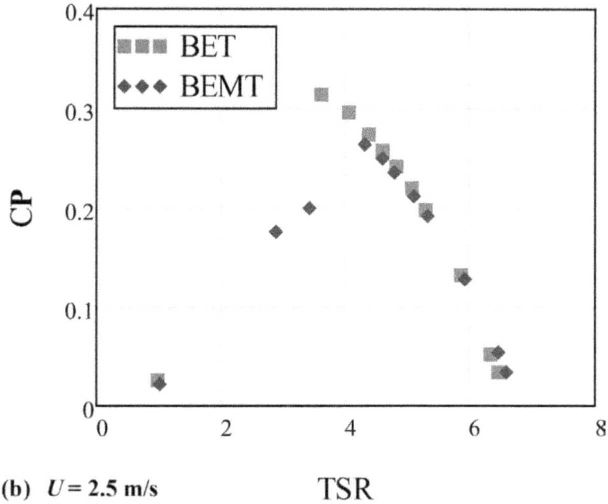

(b) *U* **= 2.5 m/s**

Figure 20: Coefficient of power vs. TSR at multiple wind speeds [39].

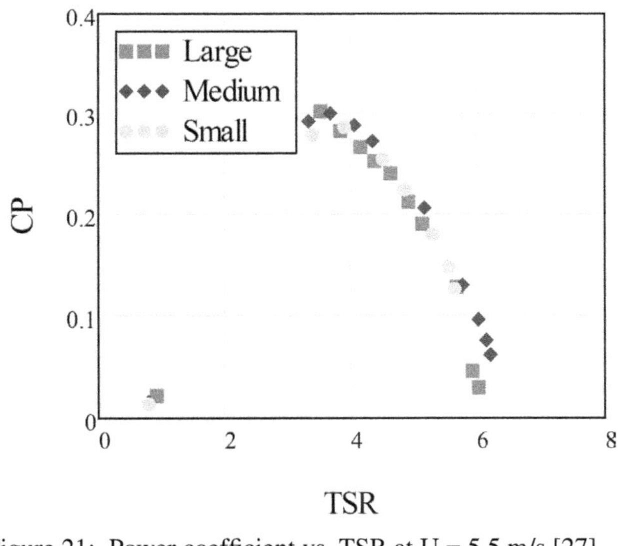

Figure 21: Power coefficient vs. TSR at U = 5.5 m/s [27].

5 Corrections for wind tunnel testing

With experimental testing of airfoils, the flow conditions in the wind tunnel are not identical to the ideal 2-D flow that an airfoil shape would experience in an open system. Because of this, multiple wind tunnel corrections have been published to account for the confinement of the wind tunnel. Barlow *et al.* provide a fairly thorough summary of wind tunnel corrections for 2-D models [78]. In the testing of six low Reynolds number airfoils, Selig *et al.* use similar correction factors but apply them in the final correction differently than Barlow *et al.* [69]. In the analysis presented in this paper, both corrections are used and compared to determine the effect of the different methods.

5.1 2-D Airfoil flow considerations

In all branches of aerodynamics that use airfoils, accurate performance lift, drag and moment coefficient data are absolutely necessary for quality design and correct analysis for predictive models. For higher velocities and subsequently higher chord-based Reynolds numbers (but not supersonic), numerical simulations such as XFOIL and PROFIL are probably adequate for use in design methods. However, for low Reynolds numbers (roughly 200,000 and below – the limit is different for every airfoil) and post-stall conditions, better predictive tools are necessary. Figure 22 shows how airfoil performance can drastically change with decreasing

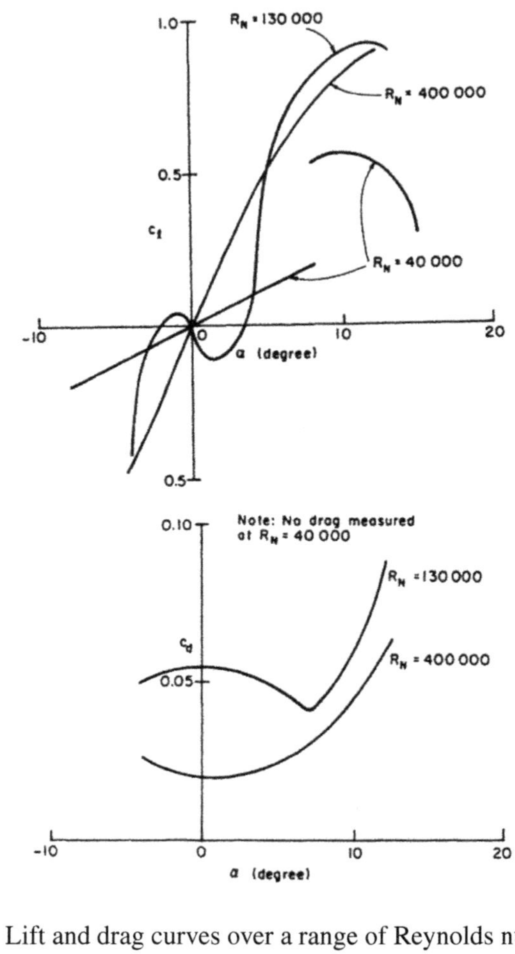

Figure 22: Lift and drag curves over a range of Reynolds numbers [78].

Reynolds numbers. In this figure, both the lift and drag coefficients (defined in eqns (6) and (7)) are plotted versus angle of attack. While numerous low-Reynolds numbers airfoils have been designed, only a fraction of those have been developed for small-scale wind turbines. Airfoils for small-scale wind turbines have unique design requirements. Some of the notable ones are structural requirements for airfoil shapes to be used at the root of the blade and the leading-edge roughness insensitivity necessary to minimise the loss of performance with deposition from bugs and other particulates [70].

All airfoils will experience low Reynolds numbers in low wind conditions or under design wind conditions near the hub where the rotational velocity is the lowest. Because of the rotational effects, the root is where the lowest relative velocities occur on the blade, thus the lowest Reynolds numbers. However, airfoil performance must be known at the specific Reynolds numbers under which the turbine blades will operate to design these blades properly. In addition, post stall data are necessary in the design of stall-regulated wind turbines, which neither PROFIL nor XFOIL can accurately predict due to flow separation and turbulent flow over most of the airfoil under these conditions.

For higher chord-based Reynolds numbers (greater than approximately 200,000) where separation bubbles are minimal and full separation only occurs at high angles of attack (greater than approximately 12°), accurate data can generally be simulated using current software like PROFIL and XFOIL. However, PROFIL does not accurately predict the drag curve for low Reynolds numbers because it does not account for the added drag due to the separation bubble as shown by McGhee *et al.* [60]. As a result, the simulated drag curve can be lower than what is experimentally measured. Additionally, the inaccurate drag curve can lead to an inaccurate prediction of the angle at which the maximum L/D occurs. XFOIL, on the other hand, was developed to account for separation bubble losses and simulates low Reynolds number flow, as shown by Drela, for the E387 airfoil [79]. Nonetheless, this code is not always accurate for other airfoils analysed at low Reynolds numbers. In particular, it does not always predict the location of turbulent separation accurately as shown by McGranahan and Selig, which can result in significant discrepancies in drag prediction [80]. Additionally, neither of these codes predicts post-stall performance accurately, which is extremely important for stall-controlled blades. Because of the limitations of the current predictive codes, taking experimental data for low Reynolds numbers is necessary to ensure accurate data are obtained for blade design.

5.2 2-D Airfoil wind tunnel corrections

Multiple corrections were applied to the data to account for the increased drag due to the airfoil mount as well as the flow conditions altered by the wind tunnel walls. Perhaps, one of the first corrections that should be considered when collecting airfoil data is the drag tare associated with the force balance. This must be accomplished with the model off the support or sting. With the model off, the wind tunnel needs to be run at the operating conditions and, depending on the type of sting, angles of attack. To account for the increased drag due to the airfoil mount, the bottom bracket of the mount (detached from the airfoil) was installed in the wind tunnel and secured at an angle of 0.0°. Subsequently, the wind tunnel was run at the test Reynolds numbers, and, using eqn (1), the measured drag was converted to a coefficient. This value was then subtracted off the measured values for drag at the corresponding Reynolds numbers. Measurements of drag should be made at the operating conditions and a table or graph for correction made. These values would then be subtracted from the total drag recorded to get the drag of the model alone. More ways to estimate drag added by the tare can be found in Barlow *et al.* [78]

The wind tunnel walls affect the 2-D flow so that it is not the same as an airfoil performing in the open air. The effects of the wind tunnel walls can be divided into three categories: blockage, streamline curvature and buoyancy (as described by Barlow *et al.* [78]).

The first type of blockage is solid blockage, ε_{sb}, which is caused by the presence of a model within the wind tunnel test section. This decreases the effective area that the air is flowing through. Applying Bernoulli's equation shows that the air velocity must increase when it flows through the decreased area, effectively increasing all aerodynamic forces and moments for that angle of attack. As seen in eqn (10), the solid blockage correction factor is merely a function of model volume, M_v and wind tunnel cross-sectional area, A_{ts}. In calculating this equation, a value of 0.74 was used forper the recommendation by Barlow *et al.* for a wing spanning the tunnel. In calculating, the simple cross-sectional area was used since the displacement thickness at the location of the model is not currently known. Notably, accounting for only the model volume does not account for the change that occurs with airfoil orientation. However, since the airfoil will be operating over a relatively small range of angles, this correction factor is sufficient

$$\varepsilon_{sb} = \frac{K_1 M_v}{A_{ts}^{3/2}}. \tag{10}$$

The second type of blockage is wake blockage, ε_{wb}, which is caused by a decreased velocity in the airfoil wake. According to the continuity equation for closed test sections, the freestream velocity must increase. The effect of the blockage is proportional to the size of the wake; thus, it is proportional to the measured value for drag. The constant of proportionality only depends on the chord of the airfoil and the height of the test section, h_{ts}, as seen in eqn (11), where c_{du} is the uncorrected drag coefficient

$$\varepsilon_{wb} = \left(\frac{c}{2h_{ts}}\right) c_{d_u}. \tag{11}$$

The second effect that is caused by the wind tunnel walls is streamline curvature. This correction occurs because the physical boundaries of the walls interfere with the natural curvature of the free air over the model, effectively increasing the airfoil effective camber and a subsequent increase in lift, moment and angle of attack. However, the drag is not affected by this phenomenon. Equation (12) documents the streamline correction factor

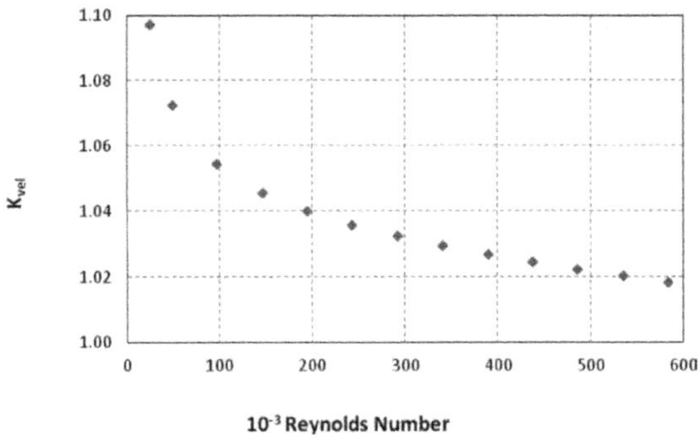

Figure 23: Buoyancy correction factor variation [76].

$$\sigma = \frac{\pi^2}{48}\left(\frac{c}{h_{ts}}\right)^2. \tag{12}$$

Buoyancy, or boundary layer growth, is the third notable effect of the wind tunnel walls. As the air moves through the wind tunnel, the boundary layer will grow increasingly large and the static pressure will decrease. Since the freestream velocity is measured upstream of the model, the actual freestream velocity at the model is higher. This correction factor must be determined experimentally for each unique wind tunnel. Equation (13) shows the form of the correction factor, and eqn (14) shows the correction factor equation used by Selig *et al.* [69]. Figure 23 shows the variation of the velocity correction factor with Reynolds number. While this correction is insignificant for Reynolds numbers above 400,000, the correction factor increases significantly as Reynolds number decreases, suggesting that such a correlation should be determined for low Reynolds number testing and should be used to correct the data.

$$K_{vel} = f(V_u), \tag{13}$$

$$K_{vel} = 1.015755 - 0.0002391V_u + 0.00001712\sqrt{V_u} + \frac{0.1296}{\sqrt{V_u}}. \tag{14}$$

Multiple methods of combining correction factors have been published, two of which will be presented and used for comparison purposes. The first is from Barlow *et al.* [78], which can be seen in eqns (15)–(19).

$$V = V_u(1+\varepsilon_b), \tag{15}$$

$$\alpha = \alpha_u + \left(\frac{1}{8\pi}\frac{c^2}{h^2+(c/4)^2} + \frac{1}{2\pi}\sigma\right)c_{l_u}, \tag{16}$$

$$c_l = c_{l_u}(1-\sigma-2\varepsilon_b), \tag{17}$$

$$c_d = c_{d_u}\left(1 - 3\varepsilon_{sb} - 2\varepsilon_{wb}\right), \tag{18}$$

$$\varepsilon_b = \varepsilon_{sb} + \varepsilon_{wb}\,. \tag{19}$$

A second method, documented by Selig *et al.* [69], can be seen in eqns (13)–(16)

$$V = V_u K_{vel}\left(1 + \varepsilon_b\right), \tag{20}$$

$$\alpha = \alpha_u + \frac{57.3\sigma}{2\pi}\left(c_{l_u} + 4c_{m_u}\right), \tag{21}$$

$$c_l = c_{l_u}\frac{(1-\sigma)}{\left(1+\varepsilon_b\right)^2}, \tag{22}$$

$$c_d = c_{d_u}\frac{1-\varepsilon_{sb}}{\left(1+\varepsilon_b\right)^2}. \tag{23}$$

Typical data with correction are shown Figure 24 [76]. In these figures, the data presented are compared with published data from UIUC.

One Reynolds number was chosen in the drag insensitive region, 200,000, as shown in Fig. 25, to validate the experimental process by comparing it with published PROFIL data, generated XFOIL data, published data for this Reynolds number, and general airfoil theory. The published experimental data were taken at UIUC. Lift and drag curves were developed from approximately –5° to 15°. The lift and drag curves from this experimental test are shown in Fig. 24. As shown, the lift curve matches PROFIL, XFOIL and published experimental data very reasonably, as well as a theoretical airfoil slope of 2π per radian or 0.1 per degree. Both XFOIL and experimental data predict stall at approximately 10°, but PROFIL data do not include a stall angle of attack. Noticeably, the UIUC drag data matched the predictions from XFOIL and PROFIL reasonably well, although the experimental data are consistently about 0.01 high in drag coefficient. Nonetheless, the drag curve follows the form of the prediction and UIUC data. The higher experimental values for drag may be due to the surface roughness of the foam airfoil model.

Figure 26 highlights the need for taking experimental data, especially at lower Reynolds numbers, as there is no other published data for this airfoil at a Reynolds number of 75,000. The data are presented for a Reynolds number of 75,000 for S823 airfoil. The XFOIL comparison does not capture the stall region for this airfoil nor does it capture the flow separation on the bottom of the airfoil at low negative angles of attack as is shown in the profile presented in Fig. 27.

After analysing the applied corrections for the data, the drag tare was the most significant correction, reducing the drag coefficient by at least 0.006 and nearly 0.01 for a Reynolds number of 75,000. The wind tunnel wall corrections, however, proved insignificant for low values of lift and drag. With a large wake due to large

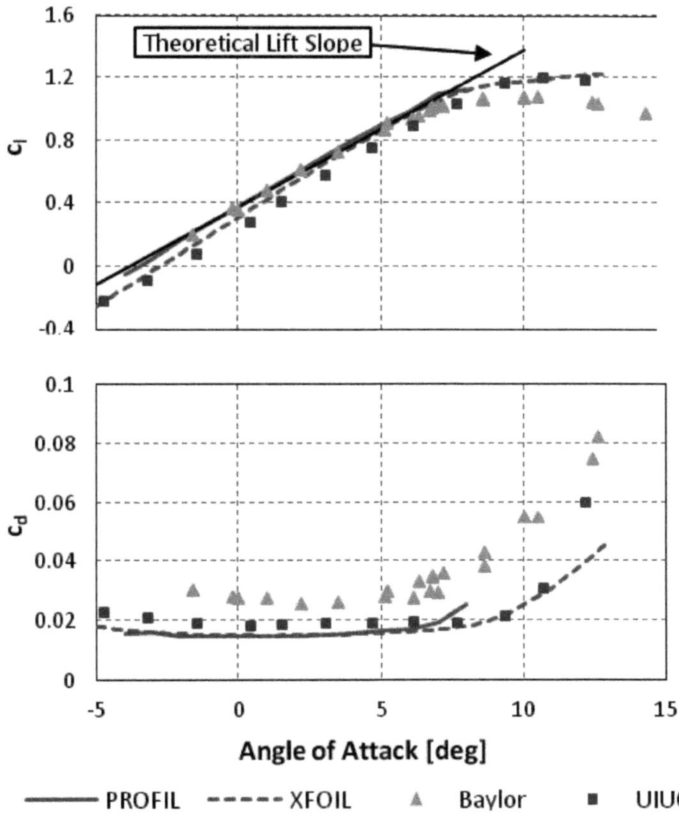

Figure 24: Lift and drag curves for an S823 airfoil at Re = 200,000 [76].

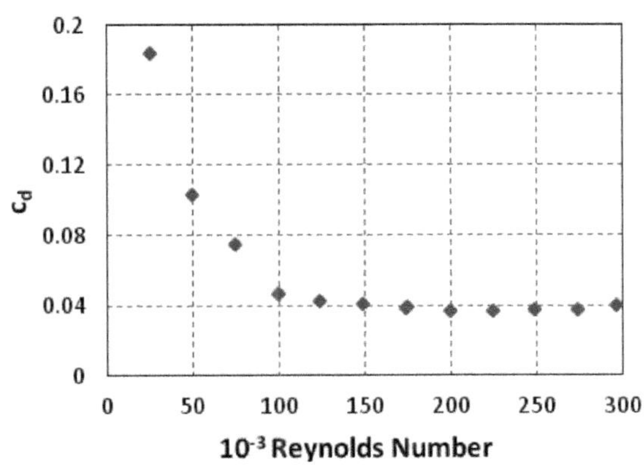

Figure 25: Drag coefficient data over a range of Reynolds numbers, angle of attack of –0.2° [76].

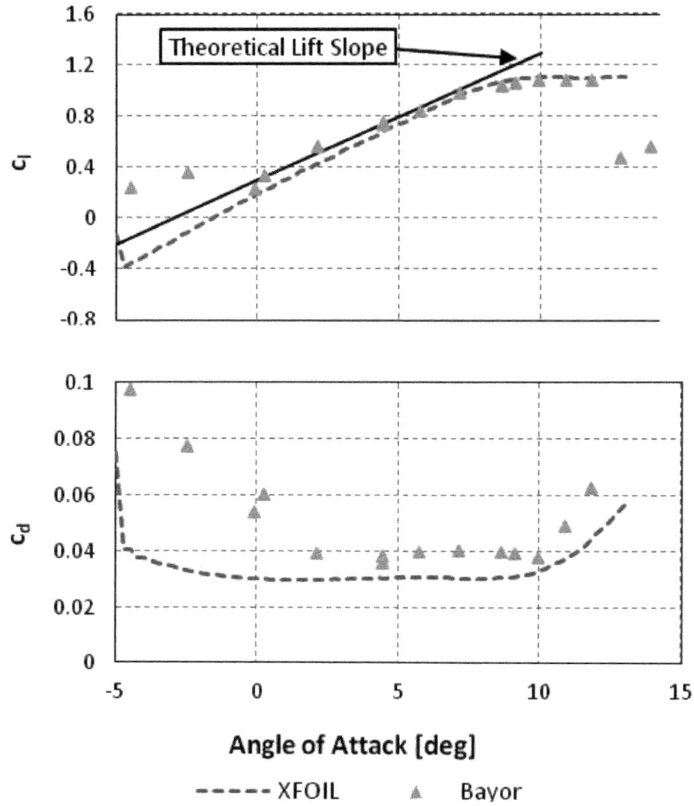

Figure 26: Lift and drag curves for an S823 airfoil at Re = 75,000 [76].

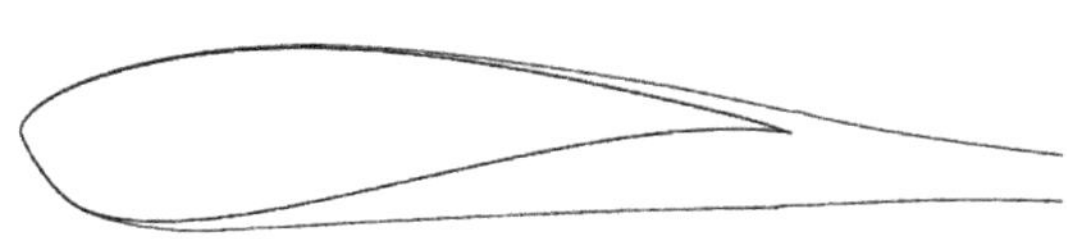

Figure 27: XFOIL streamline visualisation, Re = 75,000, $\alpha = -5°$ [76].

amounts of separation over the airfoil surface and a subsequent increase in drag, the wind tunnel wall corrections became a greater percentage of the actual value for lift and drag coefficients but were still only noticeably at higher values. By applying all of these corrections, the lift coefficient curve matched reasonably well for all the data, and the drag curve matched the data better than originally. However, the drag coefficient curve remained noticeably higher than the predictions and UIUC data, roughly 0.01 high in drag coefficient.

Figures 28–30 show the uncorrected, tared and corrected data using the methods espoused by both Selig *et al.* and Barlow *et al.* In applying the corrections to the drag data, the drag tare was applied before the wind tunnel wall corrections were

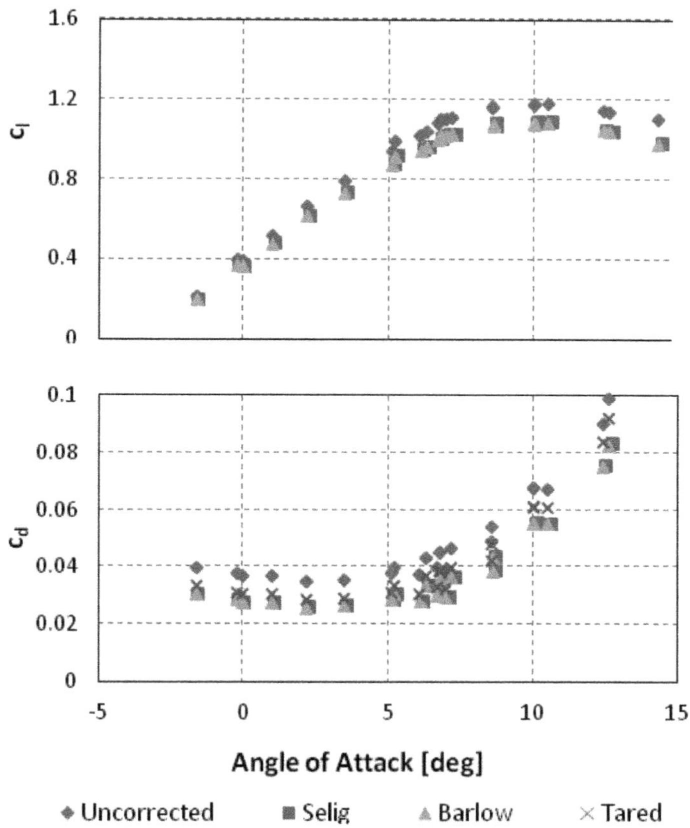

Figure 28: Corrections to an S823 airfoil at Re = 200,000 [76].

applied. These data do not include any corrections due to moment coefficient as seen in eqn (21) for the Barlow *et al.* corrections. After all the corrections, the drag curve is still noticeably higher than the predicted and UIUC data. One possibility is that the drag tare performed underestimates the actually drag added by the mount, which is somewhat likely because the shape of the curve is very similar to the other data and the lift curve matches very well. A second possibility is increased drag due to the surface roughness present on the airfoil.

Two corrections are not recommended in future studies. First, the correction for angle of attack, which never reached an order higher than the order of the fixed uncertainty of the angle measurement using both methods, is not recommended unless the angle measurement method is improved. Second, the correction for velocity has not been applied in this study because the Reynolds numbers presented are nominal and the change due to the correction is not significant enough to warrant using the corrected value. However, preliminary calculations suggest that for lower Reynolds numbers than tested in the current study, the correction may be significant and should be performed.

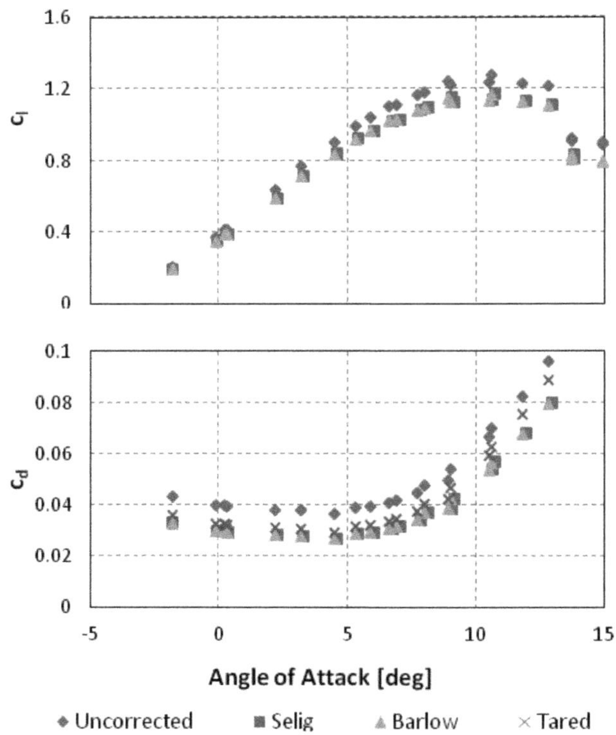

Figure 29: Corrections for an S823 airfoil at Re = 150,000 [76].

Table 3: Percent decrease due to corrections [77].

	Method	Lift		Drag	
200k	Selig *et al.*	7.15%	– 10.84%	6.93%	– 11.79%
	Barlow *et al.*	7.51%	– 11.73%	7.29%	– 12.79%
150k	Selig *et al.*	7.18%	– 10.74%	6.87%	– 11.64%
	Barlow *et al.*	7.54%	– 11.61%	7.23%	– 12.61%
75k	Selig *et al.*	7.41%	– 11.34%	7.32%	– 12.22%
	Barlow *et al.*	7.80%	– 12.32%	7.72%	– 13.31%

Comparing the two methods for wind tunnel wall corrections to each other, the method present by Barlow *et al.* is generally 0.5%–1.0% more significant than that presented by Selig *et al.* as seen in Table 3.

5.3 Blockage corrections for wind turbine rotors

The challenge of testing wind turbine blades in a small wind tunnel was explored by Van Treuren and Gregg [26]. Testing of small-scale models in wind tunnels,

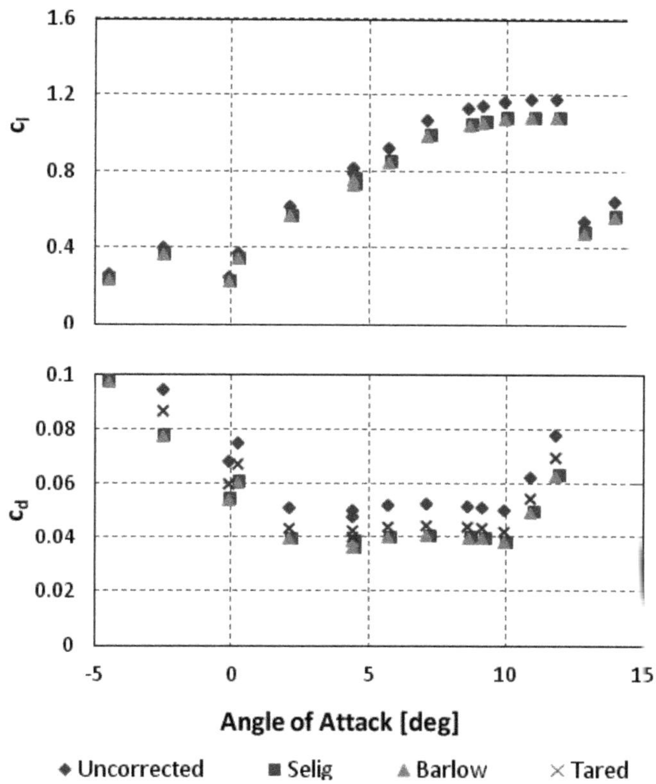

Figure 30: Corrections for an S823 airfoil at Re = 75,000 [76].

some with test sections of 1 m × 1 m or less, has been accomplished in the past [44–48, 62, 81–83]. Blockage can be a significant factor in wind tunnel testing; however, Howell *et al.* suggest that fixed tunnel blockage up to 10% does not significantly affect results [46]. Blockage due to rotating wind turbine blades has not been extensively studied but still needs to be considered. Chen and Liou have documented the correction factor necessary to account for tunnel blockage due to the rotating blades, as seen in eqn (24) [35]. The correction factor, BF, is the ratio of the freestream velocity with the turbine installed, U_M, to the freestream velocity without the turbine installed while the wind tunnel fan is operating under the same conditions, U

$$BF = \frac{U_M}{U}. \tag{24}$$

This blockage factor is used whenever a velocity is involved in calculation of a figure of merit, such as the TSR. TSR_M, shown in eqn (25), is the tip speed ratio of the model based on the velocity of the tunnel with the wind turbine in place with the wind tunnel fan operating at the speed that was found for the operation of the wind tunnel at the freestream velocity without the wind turbine present in the test section. Using the blockage factor, eqn (26) calculates the actual TSR for the wind turbine power system

$$TSR_M = \frac{\Omega R}{U_M}, \tag{25}$$

$$TSR = TSR_M \mathrm{BF} = \frac{\Omega R}{U}. \tag{26}$$

This blockage factor, BF, is applied to the coefficient of power based on the U_M ($C_{P,M}$, eqn (27)) and used to calculate the corrected power coefficient, C_P, as seen in eqn (28). For all the power coefficients, P is the rotor power from the wind turbine (a function of the angular velocity Ω), ρ is the air density and A is the area swept by the rotating wind turbine blades

$$C_{P,M} = \frac{P_{Rotor}}{P_{Freestream,M}} = \frac{\tau\Omega}{\frac{1}{2}\rho\pi \mathrm{R}^2 U_M^3}, \tag{27}$$

$$C_P = \frac{P_{Rotor}}{P_{Freestream}} = C_{P,M} BF^3 \frac{\tau\Omega}{\frac{1}{2}\rho\pi \mathrm{R}^2 U^3}. \tag{28}$$

Figure 21 illustrates the application of these corrections to wind tunnel system testing. These wind tunnel blades, as seen in Fig. 16a, have diameters of 0.5 m. 0.4 m and 0.3 m tested in the wind tunnel test section described in Fig. 15. Wind tunnel blockage for these diameters is 52.8%, 33.8% and 19.0%, respectively. This is well past the upper limit of 10% that most tests seek to be below as lower than 10% typically neglects any blockage corrections.

5.4 3-D flow effects

Additionally, 3-D flow effects, not taken into account in traditional wind turbine design using 2-D airfoil data, have been shown to influence aerodynamic efficiency. Robinson *et al.* examined the impact of three-dimensionality, unsteadiness, and flow separation on downwind horizontal axis wind turbines (HAWTs) [84]. Their results show that '... quasi-static separation and dynamic stall analysis methods relying on purely 2-D flow characterisations may not be capable of simulating the complex 3-D flows observed with these data'. These experiments used a 10-m diameter rotor with constant chord blades and an S809 airfoil tested at several different pitch angles. They found aerodynamic loads were far in excess of 2-D wind tunnel data for near- and post-stall conditions. Testing of airfoil data and comparison with actual wind tunnel data were also done by Butterfield *et al.* [42]. They confirm that stall-regulated HAWTs show peak power levels significantly greater than predicted by BEMT using 2-D airfoil data. They warn that 'wind turbine designers should be aware of the differences when designing stall-controlled wind turbine blades or they may underestimate the peak performance and loads'. They also suggest more research is needed in these areas.

Using CFD analysis, Qiu *et al.* showed that there is significant spanwise flow on the suction surface of a rotating wind turbine blade [85]. The control of these behaviours using roughness has begun to be investigated and has shown to have a significant effect either to delay separation [86,87] or to increase drag [88–90]. Thus, the effects of roughness must be further analysed to determine what designs could be improved by the application of intentional roughness. At the low Reynolds numbers, roughness was found to improve performance significantly.

NOMENCLATURE

A	Wind turbine area swept by rotor blades (amps)
A_{ts}	Wind tunnel test section area minus the displacement thickness around the perimeter
a	Axial momentum factor
a'	Tangential momentum factor
BEMT	Blade element momentum theory
BET	Blade element theory
BF	Blockage factor
C, c	Chord
C_D, c_d	Drag coefficient
C_{d_u}	Uncorrected drag coefficient
C_L, c_l	Lift coefficient
C_{l_u}	Uncorrected lift coefficient
C_M, c_m	Moment coefficient
C_{m_u}	Uncorrected moment coefficient
C_P	Rotor power coefficient
$C_{P,M}$	Rotor power coefficient with wind turbine in the tunnel (blockage)
$C_{P,S}$	System power coefficient
C_T	Thrust coefficient
D	Drag
F_c	Centrifugal forces
h_{ts}	Wind tunnel test section height
J	Advance ratio
K_1	Constant in solid blockage equation
K_p	Rotor speed power coefficient
K_{vel}	Velocity correction factor
L	Lift
l	Unit length

M, M_A	Pitching moment and aerodynamic moment
M_v	Airfoil model volume
P, P_{Rotor}	Power androtor power
$P_{Freestream}$	Freestream power
$P_{Freestream,\ m}$	Freestream power (blockage)
p_∞	Freestream static pressure
p_d	Static pressure at the actuator disk
Q, τ	Torque
R, r, R_N	Radius
Re, R_N	Reynolds number based on chord
RPM	Revolutions per minute
sc	Scaling parameter
T	Thrust
TSR_M	Tip Speed Ratio with wind turbine in the tunnel (blockage)
U, U_∞	Freestream velocity
U_d	Velocity at the actuator disk
U_M	Wind tunnel velocity with the wind turbine in the tunnel (blockage)
U_{rel}	Incoming velocity to the turbine blade (axial and rotational combined)
U_w	Wake velocity
V	Freestream velocity
V_u	Uncorrected freestream velocity
W	Weight
x	Chordwise distance
y	Vertical distance
a	Angle of attack
a_u	Uncorrected angle of attack
β	Pitch angle
δr	Segment of radial length
ε_b	Blockage correction factor
ε_{sb}	Solid blockage correction factor
ε_{wb}	Wake blockage correction factor
μ_{air}	Dynamic viscosity of air
σ	Streamline curvature correction factor
σ_g	Gravitational stress
σ_A	Aerodynamic stress
σ_C	Centrifugal stress
υ	Translation velocity and kinematic viscosity
λ, TSR	Tip speed ratio
ρ, ρ_{air}	Air density
ϕ	Flow angle
ω	Natural frequencies
Ω	Blade rotational speed

References

[1] Spera, D.A., ed., *Wind Turbine Technology: Fundamentals Concepts in Wind Turbine Engineering*. ASME Press: New York, NY, 2009.

[2] Manwell, J.F., McGowan, J.G. & Rogers, A.L., *Wind Energy Explained: Theory, Design and Application*, 2nd edn. John Wiley and Sons, Ltd.: West Sussex, UK, 2009.

[3] Burton, T., Sharpe, D., Jenkins, N. & Bossanyi, E., *Wind Energy Handbook*. John Wiley and Sons, Ltd.: West Sussex, UK, 2008.

[4] Rankine, W.J.J., On the mechanical principles of the action of propellers. *Transactions of the Institute of Naval Architects*, **6**, pp. 13–30, 1865.

[5] Froude, R.E., On the elementary relation between pitch, ship, and propulsive efficiency. *Transactions of the Institute of Naval Architects*, **10**, pp. 47–57, 1878.

[6] Joukowski, N.E., Travanx du Bureau des Calculs et Essais Aeronautiques de l'Ecole Supericure Technique de Moscou, 1918.

[7] Betz, A., Schraubenpropeller mit geringstem Energieverlust, Gottinger Nachrichten, Mathematisch-Physikalische Klasse, pp. 193–213, 1919.

[8] Drzewiecki, S., Sur une methode pour la determination des elements mechaniques des propulseurs helicoidaux. *Comptes Rendus de l'Academie des Sciences, Paris*, **114**, pp. 820–822, 1982.

[9] Glauert, H., Airplane Propellers. *Aerodynamic Theory,* Vol 4, Div. 1, ed. W.F. Durand, Julius Springer: Berlin, Germany, pp 169–360, 1935.

[10] Buhl Jr., M.L., *WT_Perf User's Guide*, National Wind Technology Center, National Renewable Energy Laboratory, 2004.

[11] Burdett, T.A., Small-scale HAWT design process for Waco, ME 5396 Wind Energy Design Project. Spring 2011.

[12] Blanch, M.J., Reynolds number effects on the stall of a small horizontal axis wind turbine. *Wind Energy Conversion 1996*. Mechanical Engineering Publications Limited: London, 1997.

[13] Simms, D., Schreck, S., Hand, M. & Fingersh, L.J., NREL unsteady aerodynamics experiment in the NASA-Ames wind tunnel: a comparison of predictions to measurements, NREL/TP-500-29494, 2001.

[14] Sicot, C., Devianant, P., Loyer, S. & Hureau, J., Rotational and turbulence effects on a wind turbine blade: investigation of the stall mechanisms. *Journal of Wind Engineering and Industrial Aerodynamics*, **96**, pp. 1320–1331, 2008.

[15] Bowen, A., Huskey, A., Link, H., Sinclair, K., Forsyth, T., van Dam, J. & Smith, J., Small wind turbine testing results from the national renewable energy laboratory, NREL/CP-500-48089, Presented at the *International Small Wind Conference*, Glasgow, Scotland, 2010.

[16] Gregg, J.R., Burdett, T.A., Van Treuren, K.W. & McClain, S.T., Design considerations, performance enhancing techniques, and wind tunnel testing for

small scale, low Reynolds number wind turbines. IMECE2011-65351, *ASME International Mechanical Engineering Congress & Exposition*, Denver, CO, November 11–17, 2011.

[17] Evangelista, R., McGhee, R.J. & Walker, B.S., Correlation of theory to wind-tunnel data at Reynolds numbers below 500,000, *Low Reynolds Numbers Aerodynamics Conference*, Notre Dame, IN, June 5–7, 1989.

[18] Hau, E., *Wind Turbines: Fundamentals, Technologies, Applications, Economics*. Springer-Verlag: Berlin, Germany, 2006.

[19] Tangler, J.L. & Somers, D.M., NREL Airfoil Families for HAWTs, *American Wind Energy Association WindPower '95 Conference,* Washington, DC, March 26-30, 1995.

[20] Jackson, K.L., Scaling wind turbine fatigue design loads. *Wind Energy*, **15**, pp. 189–196, 1994.

[21] Srikanth, N. & Funk, J.L., Geometric scaling and long-run reductions in cost:the case of wind turbines. *2011 IEEE International Technology Management Conference (ITMC)*, San Jose, CA, June 27–30, 2011, pp. 691–696, 2011.

[22] Moe, G., What is the optimum size for a wind turbine? OMAE2007-29035, *26th International Conference on Offshore Mechanics and Arctic Engineering*, San Diego, CA, June 10–15, 2007, pp. 409–416, 2007.

[23] Twele, R. & Gasch, R., *Wind Power Plants: Fundamental, Design, Construction, and Operation*. Solarpraxis: Berlin, Germany, 2002.

[24] Petersen, H., The scaling laws applied to wind turbine design. *Wind Engineering*, **8(2)**, 1984, pp. 99–108, 1984.

[25] Shrestha, G., Polinder, H. & Ferreria, J.A., Scaling laws for direct drive generators in wind turbines. *2009 IEEE International Electric Machines and Drives Conference (IEMDC)*, Miami, FL, May 3–6, 2009.

[26] Van Treuren, K.W. & Gregg, J.R., Testing rotating horizontal axis wind turbine blade designs in a laboratory wind tunnel. GT2010-23575, *Proceedings of the ASME Turbo Expo*, Glasgow, UK, 2010.

[27] Burdett, T.A. & Van Treuren, K.W., Scaling Small Scale Wind Turbines for Wind Tunnel Testing, GT2012-68359, Proceedings of the ASME IGTI Turbo Expo, Copenhagen, Denmark, June 11–15, 2012.

[28] Wainauski, H.S., Wind tunnel tests of a 1/30 scale model of a 3.5 megawatt wind turbine. *Proceedings of the 14th Intersociety Energy Conversion Engineering Conference*, Boston, MA, August 5–10, 1979.

[29] Gao, X. & Hu, J., Numerical research of Reynolds number impact on scale model of wind turbine. *IEEE World Non-Grid-Connected Wind Power and Energy Conference*, Nanjing, China, pp. 1–4, 2009.

[30] Burdett, T., Gregg, J. & Van Treuren, K., An examination of the effect of Reynolds number on airfoil performance. *Proceedings of ASME 2011 5th International Conference on Energy Sustainability*, Washington, DC, August 7–10, 2011.

[31] Snel, H., Schepers, J.G. & Montgomerie, B., The MEXICO project (Model Experiments in Controlled Conditions): the database and first results of data

processing and interpretation. *Journal of Physics: Conference Series* **75**, 012014, 2007.

[32] Ronsten, G., Static pressure measurements in a rotating and non-rotating 2.375m wind turbine blade. Comparison with 2-D calculations. *Journal of Wind Engineering and Industrial Aerodynamics,* **39**(1–3), pp. 105–118, 1992.

[33] Snel, H., Houwindk, R., Bosschers, R., Piers, W.J., Van Bussel, G.J.W. & Bruining, A., Sectional prediction of 3-D effects for stalled flow on rotating blades and comparison with measurements. *Proceedings of the European Community Wind Energy Conference*, Germany, March 8–12, 1993.

[34] Cho, T., Chealwan K. & Duckjoo, L., Acoustic measurements for 12% scaled model of NREL Phase VI wind turbine by using beamforming. *Elsevier Current Applied Physics*, **10**, pp. 320–325, 2010.

[35] Chen, T.Y. & Liou, L.R., Blockage corrections in wind tunnel tests of small horizontal-axis wind turbines. *Elsevier Experimental Thermal and Fluid Science,* **35**, pp. 365–569, 2011.

[36] Hulskamp, A.W., Van Wingerden, J.W., Barlas, T., Champliaud, H., van Kuik, G.A.M., Bersee, H.E.N. & Verhaegen, M., Design of a scaled wind turbine with a smart rotor for dynamic load control experiments. *Journal of Wind Energy,* **14(3)**, pp. 339–354, April 2011.

[37] Lissaman, P.B.S., Low-Reynolds-number airfoils. *Annual Review of Fluid Mechanics*, **15**, pp. 223–239, 1983.

[38] Burdett, T.A., Gregg, J.R. & Van Treuren, K.W., A numerical study on improving airfoil performance at low Reynolds numbers for small-scale wind turbines using intentional roughness. *Proceedings of 50th AIAA Aerospace Sciences Meeting*, Nashville, January 9–12, 2012.

[39] Burdett, T.A. & Van Treuren, K.W., 2012, "A Theoretical and Experimental Comparison of Optimized BET and BEMT Blade Designs," GT2012-68350 Proceedings of ASME Turbo Expo, Copenhagen, Denmark, June 11–15 2012

[40] http://www. nrel. gov/wind/ (accessed 10 December 2011).

[41] Maeda, T., Ismaili, E., Kawabuchi, H. & Kamada, Y., Surface pressure distribution on a blade of a 10 m diameter HAWT (field measurements versus wind tunnel measurements). *Journal of Solar Energy Engineering*, **127**, pp. 185–191, 2005.

[42] Butterfield, C.P., Scott, G. & Musial, W., Comparison of wind tunnel airfoil performance data with wind turbine blade data. *Journal of Solar Energy Engineering*, **114**, pp. 119–124, 1992.

[43] Gregg, J.R., Merchant, J.S., Van Treuren, K.W. & Gravagne, I.A., Experimental analysis of a counter-rotating wind turbine. IMECE2009 – 11355, *Proceedings of the International Mechanical Engineering Congress and Exposition*, Lake Buena Vista, FL, November 13–19, 2009.

[44] Rector, M.C. & Visser, K.D., Aerodynamic design of a small contra-rotating HAWT. *45th AIAA Aerospace Sciences Meeting*, Reno, NV, January 8–11, 2007.

[45] Ushiyama, I., Shimota, T. & Miura, Y., An experimental study of the two-staged wind turbines. *WREC*, pp. 909–912, 1996.
[46] Howell, R., Qin, N., Edwards, J. & Durrani, N., Wind tunnel and numerical study of a small vertical axis wind turbine. *Renewable Energy*, **35(2)**, pp. 412–422, 2010.
[47] Pawsey, N.C.K. & Barratt, A.J., Evaluation of a variable-pitch vertical axis wind turbine. *Wind Engineering*, **23(1)**, pp. 23–30, 1999.
[48] Georgescu, A., Georgescu, S. & Degeratu, M., Experimental setup designed for testing a cross-flow water turbine in a wind tunnel. *U. P. B. Sci. Bull., Series D, Vol. 70, Issue 4*, 5th National Conference of Romanian Hydropower Engineers, Dorin Pavel, May 22–23, 2008, Bucharest, Romania, pp. 191–200, 2008.
[49] Deam, R.T., On scaling down turbines to millimeter size. *Journal of Engineering for Gas Turbines and Power*, **130**, pp. 52301-1–52301-9, 2008.
[50] Jacobs, E.N., Ward K.E. & Pinkerton, R.M., The characteristics of 78 related airfoil sections from tests in the variable-density wind tunnel, NACA-TR-460, 1993.
[51] Abbott, I.H., Von Doenhoff, A.E. & Stivers, L., Summary of airfoil data, NACA-TR-824, 1945.
[52] NREL home page. http://www. nrel. gov/wind/pubs_research. html (accessed 14 November 2009).
[53] Giguere, P. & Selig, M.S., Low Reynolds number airfoils for small horizontal wind turbines. *Wind Engineering*, **21(6)**, pp. 367–380, 1997.
[54] Fuglsang, P., Bak, C., Gaunaa, M. & Antoniou, I., Design and verification of the Ris0-B1 airfoil family for wind turbines. *Transactions of the ASME*, **126**, pp. 1002–1010, 2004.
[55] Sommers, D.M., The S816, S817, and S818 Airfoils, NREL/SR-500-36333, December 2004.
[56] Sommers, D.M., The S827, and S828 Airfoils, NREL/SR-500-36343, January 2005.
[57] Bak, C. & Fuglsang, P., Modification of the NACA 632-415 leading edge for better aerodynamic performance. *Journal of Solar Engineering*, **124**, pp. 327–334, 2002.
[58] Albertani, R., Wind-tunnel study of gurney flaps applied to micro aerial vehicle wing. *AIAA Journal*, **46(6)**, pp. 1560–1562, 2008.
[59] Mcghee, R.J. & Beasley, W.D., Wind-tunnel results for a modified 17-percent-thick low-speed airfoil section, NASA-TP-1919, 1981.
[60] McGhee, R., Walker, B. & Millard, B., Experimental results for the Eppler 387 airfoil at low reynolds number in the langley low-turbulence pressure tunnel, NASA Technical Memorandum 4062, 1988.
[61] Hu, H. & Yang, Z., An experimental study of the laminar flow separation on a low-Reynolds-number airfoil. *Journal of Fluids Engineering*, **130**, pp. 1–11, 2008.
[62] Hirahara, H., Hossain, M., Kawahashi, M. & Nonomura, Y., Testing basic performance of a very small wind turbine designed for multi-purposes. *Renewable Energy*, **30(8)**, pp. 1279–1297, 2005.

[63] Gross, A., Fasel, H.F., Freiderich, T. & Kloker, M.J., Numerical investigation of S822 wind turbine airfoil. *40th AIAA Fluid Dynamics Conference and Exhibit*, Chicago, IL, 2010.

[64] Giguere, P. & Selig, M., Low Reynolds number airfoils for small horizontal axis wind turbines. *Wind Engineering*, **21(6)**, pp. 367–380, 1997.

[65] Selig, M., Low Reynolds number airfoil design lecture notes. *VKI Lecture Series*, NATO Research and Technology Organization, Applied Vehicle Technology Panel, 2003.

[66] Sicot, C., Devianant, P., Loyer, S. & Hureau, J., Rotational and turbulence effects on a wind turbine blade: investigation of the stall mechanisms. *Journal of Wind Engineering and Industrial Aerodynamics*, **96**, pp. 1320–1331, 2008.

[67] Watanabe, Y., Iwashita, H. & Ito, M., Shape optimum design of horizontal axis wind turbine in low Reynolds number range. *European Wind Energy Conference*, Milan, Italy, 2007.

[68] Greco, L., Testa, C. & Salvatore, F., Design oriented aerodynamic modelling of wind turbine performance. *Journal of Physics: Conference Series*, **75**, pp. 1–12, 2007.

[69] Selig, M.S. & McGranahan, B.D., Wind tunnel aerodynamic tests of six airfoils for use on small wind turbines. *ASME Journal of Solar Energy Engineering*, **126**, pp. 986–1001, 2004.

[70] Somers, D.M., The S833, S834, and S835 airfoils. NREL/SR-500-36340, National Renewable Energy Laboratory, 2005.

[71] Somers, D.M., The S822 and S823 airfoils. NREL/SR-500-36333. National Renewable Energy Laboratory, 2005.

[72] Selig, M.S. & McGranahan, B.D., Wind tunnel aerodynamic tests of six airfoils for use on small wind turbines. *Journal of Solar Energy Engineering*, **126**, pp. 986–1001, 2004.

[73] Selig, M.S., Guglielmo, J.J., Broeren, A.P. & Giguère, P., *Summary of Low-Speed Airfoil Data*, Vol. 1. SoarTech Publications: Virginia Beach, VA, 1995.

[74] Selig, M.S., Lyon, C.A., Giguère, P., Ninham, C.N. & Guglielmo, J.J., *Summary of Low-Speed Airfoil Data*, Vol. 2. SoarTech Publications: Virginia Beach, VA, 1996.

[75] Lyon, C.A., Broeren, A.P., Giguère, P., Gopalarathnam, A. & Selig, M.S., *Summary of Low-Speed Airfoil Data*, Vol. 3. SoarTech Publications: Virginia Beach, VA, 1998.

[76] Burdett, T., Applying wind tunnel corrections to S823 airfoil data. Paper written for *Experimental Methods for Heat Transfer and Fluid Flow Class* at Baylor University, May 9, 2011.

[77] Gregg, J.R. & Van Treuren, K.W., 2010, "Experimental Testing of Periodic Roughness Elements on a Small Scale Wind Turbine Blade," IMECE2010-38863, Proceedings of the *ASME International Mechanical Engineering Congress & Exposition*, British Columbia, Canada, November 12–18, 2010.

[78] Barlow, J.B., Rae Jr., W.H. & Pope, A., *Low-Speed Wind Tunnel Testing*, 3rd edn. John Wiley and Sons, Inc.: New York, pp. 349–362, 673–674, 1999.

[79] Drela, M., XFOIL: an analysis and design system for low Reynolds number airfoils. *Conference on Low Reynolds Number Airfoil Aerodynamics*, University of Notre Dame, 1989.

[80] McGranahan, B.D. & Selig, M.S., Surface oil flow measurements on several airfoils at low Reynolds numbers. AIAA 2003-4067, *21st AIAA Applied Aerodynamics Conference*, Orlando, FL, June 23–26, 2003.

[81] Gregg, J.R., Merchant, J.S., Van Treuren, K.W. & Gravagne, I.A., Experimental analysis of a counter-rotating wind turbine. IMECE2009-11355, *Proceedings of the International Mechanical Engineering Congress and Exposition*, Lake Buena Vista, FL, 2009.

[82] Baansal, A., Howey, D. & Holmes, A., CM-scale air turbine and generator for energy harvesting from low-speed flows. *Transducers*, 2009, Denver, CO, pp. 529–532, 2009.

[83] Predescu, M., Bejinariu, A., Nedelcu, A., Mitroi, O., Nae, C., Pricop, M. & Craciunescu, A., Wind tunnel assessment of small direct drive wind turbines with permanent magnet synchronous generator. ICREPQ 2008, Santander, Spain, 2008.

[84] Robinson, M.C., Hand, M.M., Simms, D.A. & Schreck, S.J., Horizontal axis wind turbine aerodynamics: three-dimensional, unsteady, and separated flow influences. NREL/CP-500-26337, *Presented at the 3rd ASME/JSME Joint Fluids Engineering Conference*, San Francisco, CA, 1999.

[85] Qiu, M., Zhou, Z. & Zhang, J. Three-dimensional flow simulation for horizontal axis wind turbine. *2009 World Non-Grid-Connected Wind Power and Energy Conference*, IEEE, Nanjing, China, 2009.

[86] Viswanath, P.R., Aircraft viscous drag reduction using riblets, *Progress in Aerospace Sciences*, **38**, pp. 571–600, 2002.

[87] Oliver, A. & Wootton, L.R., Use of air jet vortex generators leading to improved energy capture and power regulation of a wind turbine. *Wind Energy Conversion 1996*, Mechanical Engineering Publications Limited: London, 1997.

[88] Khalfallah, M.G. & Koliub, A.M. Effect of dust on the performance of wind turbines. *Desalination*, **209**, pp. 209–220, 2007.

[89] Pechlivanoglou, G., Fuehr, S., Nayeri, C. & Paschereit, C., The effect of distributed roughness on the power performance of wind turbines. GT2010-23258, *Proceedings of the ASME Turbo Expo*, Glasgow, UK, 2010.

[90] Ren, N. & Ou, J., Numerical simulation of surface roughness effect on wind turbine thick airfoils. *IEEE*, 978-1-4244-2487-0, 2009.

CHAPTER 5

Computational Fluid Dynamics Approach for Wind Turbine Blade Aerodynamics Design

R.J. Malloy[1] & R.S. Amano[2]
[1]*Milwaukee Electric Tool Corp.*
[2]*Department of Mechanical Engineering, University of Wisconsin-Milwaukee*

Abstract

The blades of a wind turbine are a significant component of the structure and therefore much attention should be given to their design. They are the elements of a wind turbine responsible for extracting energy from the moving air by virtue of their shape. How effective the blade shape is at doing this will affect the overall efficiency and performance of the unit, which are of great interest to manufactures and their customers alike. Since blades of commercial wind turbines can reach upwards of 20 m in length, physical test of various designs can become expensive and time consuming. Although there are some general analytical formulas and procedures [1] for determining blade shapes, these tend fall short of producing optimal designs given the several significant in accurate assumptions their derivation is founded upon. It is for this reason, researches have turned to computer simulation to develop and optimize wind turbine blade shape. Computational fluid dynamics, or CFD, is a powerful tool that has been leveraged in many industries including the wind power industry. When applied and executed correctly simulation results very closely approximate actual blade performance and thus can be used to evaluate and identify optimal designs that will perform up to expectations in the real world.
Keywords: CFD, finite-volume method, HAWT, wind turbine blade design.

1 Introduction

Blade element moment (BEM) method was used to construct a straight edge blade prototype whose optimal oncoming wind and rotation speeds were 7 m/s and 20 rpm. The blade has a length of 20 m and uses a constant airfoil cross-section NACA 4412. To test this new design, the performance of both blades will be measured using computational fluid dynamics (CFD) at a wind speed of 10 m/s. In addition to this straight blade, a swept edge blade was also tested. This swept edge blade has the same characteristics as the straight edge except for the trajectory of the edge. Each cross section has the same dimensions at the same distance from the hub as its corresponding section in the straight edge blade.

The reason for testing this new geometry is that the straight edge blade is constructed using a formulaic approach, which treats the airflow over the blade as perfectly perpendicular to the leading edge and neglects any spanwise component. The swept edge blade profile aims to accommodate the spanwise velocity component and delay the stall point of the rotor. This geometry has largely been uninvestigated using the CFD approach. Recently, an investigation into the loading and dynamic behaviour of a swept blade was published by Larwood and Zuteck [2]. They used codes developed by the National Renewable Energy Laboratories (NREL), which used a more analytical approach. A CFD approach is more suitable for this investigation since it is purely an aerodynamic study and CFD yields very accurate results that are quantitative as well as qualitative.

Examples of CFD's capability of producing accurate qualitative as well as quantitative results for wind turbine analysis are abundant in scientific literature for the past 20 years. One example is the study was performed by Ferre and Munduate [3]. In this instance, CFD was used to analyse the flow conditions for three tip designs shown in Fig. 1.

The simulation was done for attached flow conditions only. Because of this, an unstructured mesh was used since it is quicker and easier than a structured mesh and high boundary layer resolution is not required. To take into account the effect of the viscous boundary layer, a wall function could be utilized although it was stated that resolution of the viscous boundary layer with a fine mesh is preferred. A moving reference frame with periodic boundary conditions was used to model the rotation of the blade, whereas the k–ω SST turbulence closure model [4] was used to model the turbulent flow. Results show good agreement with experimental data in predicting the local pressure coefficient and the normal force coefficient at locations on the blade under investigation (Fig. 2).

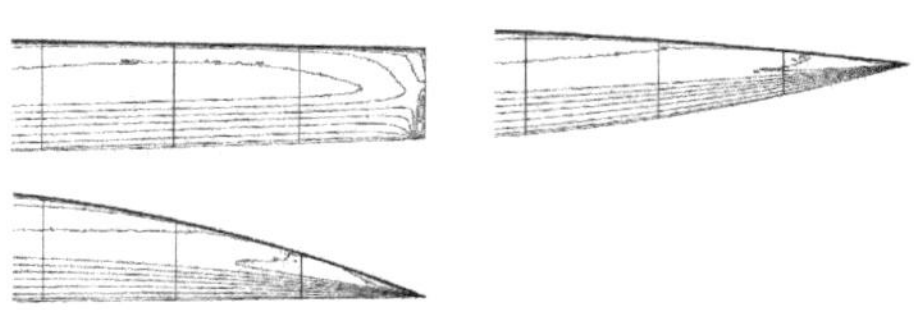

Figure 1: Pressure profiles on various tip geometries.

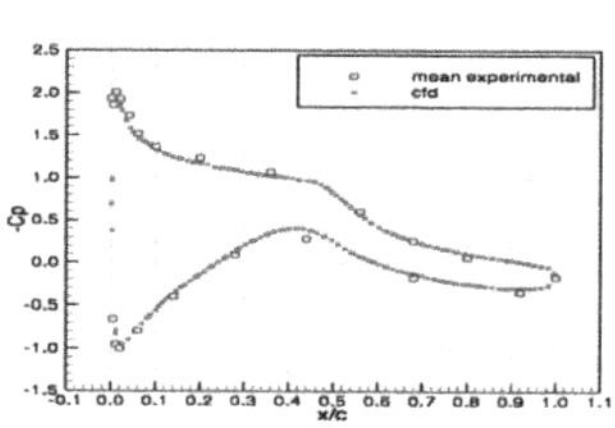

Figure 2: Pressure distribution for 63% radial location for the VI Blade.

2 Numerical Simulation Approach to Wind Turbine Blade Performance Evaluation

As one can imagine, it is critical to understand the performance characteristics associated with a wind turbine blade shape before the expensive and time-consuming processes of fabrication and installation are undertaken. There are mainly two methods used to accomplish this. The first is an analytical approach that is based on the BEM theory. This approach breaks the blade up into a finite number of elements perpendicular to span of the blade as shown in Fig. 3.

Each element represents a 2-D cross section of the blade whose performance can be evaluated using lift and drag data gathered from wind tunnel testing of the section's shape. The total torque produced by a blade for a given oncoming wind speed can be approximated by the sum of the torques produced by each section. This approach is efficient in that it does not require extensive computational resources and generally produces good results for traditional blade shapes in flow conditions where there is little or no flow separation at the trailing edge of the blade. When a wind turbine blade is in a stall condition, unlike that of an aircraft, a spanwise wind velocity component results from its rotational motion. This velocity component produces a favourable pressure gradient that holds the flow attached to the blade resulting in higher lift values at higher angles of attack than predicted by standard wind tunnel tests of the same airfoil cross sectional shape. This phenomenon is not accounted for using the BEM theory, which assumes no spanwise velocity component. It is for this reason that CFD analysis is necessary to evaluate new blade designs. Thus, for more complex blade shapes and flow conditions a numerical simulation is required.

3 Numerical Methods

The governing equations for viscous, incompressible flow can be described by the Navier–Stokes equations as stated in eqn (1). Due to their mixed elliptic–parabolic

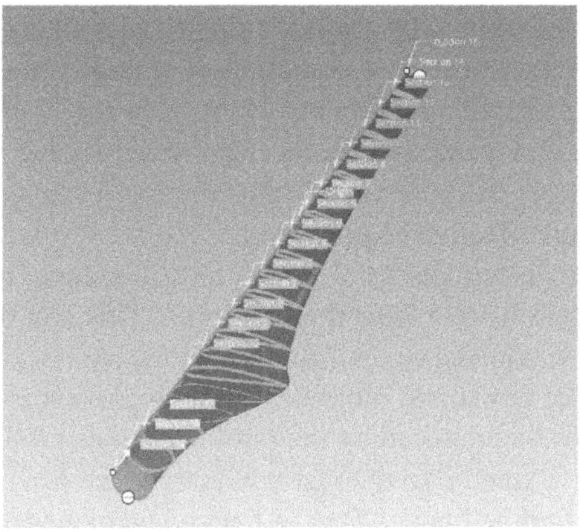

Figure 3: Airfoil cross sections of a wind turbine blade.

behavior, a pressure correction technique is widely used to solve them. This technique is employed in the SIMPLE algorithm, which couples the pressure and velocity. This method was used in this investigation along with the first-order upwind discretization scheme for all equations. All solutions were run till they were fully converged and all converged below an order of 1×10^4.Traditionally, most initial designs and performance predictions of wind turbine blades have been based on the BEM theory. This approach uses 2-D airfoil data to determine the aerodynamic loads experienced by the blades analytically. This method is fast and yields a reasonable prediction for attached flow conditions. However, the BEM theory tends to underpredict power extracted during the stall condition [5]. When a wind turbine blade is in a stall condition, unlike that of an aircraft, a spanwise wind velocity component results from its rotational motion. This velocity component produces a favourable pressure gradient, which holds the flow attached to the blade resulting in higher lift values at higher angles of attack than predicted by standard wind tunnel tests of the same airfoil cross-sectional shape. This phenomenon is not accounted for using the BEM theory, which assumes no spanwise velocity component. It is for this reason that CFD analysis is necessary to evaluate new blade designs. CFD does not use predetermined airfoil data to predict the blade performance but instead solves the governing fluid flow equations at thousands of positions on and around the blade in an iterative process. This approach allows the model to take into account any spanwise wind velocity component, which BEM theory cannot.

Numerical simulation is accomplished by breaking up the flow domain into a finite number of elements resulting in a mesh and solving the governing fluid flow equations in an iterative process from element to element. This process will yield values of velocity and pressure at every element after being determined from the prescribed boundary conditions. This approach is widely used in many applications including wind turbine performance evaluation.

A typical wind turbine analysis begins, as with any analysis, with the construction of the flow domain. This is generally the most time-consuming and challenging aspect of a CFD study because of the length scales involved. The elements near the surface of the blade need to be small in order to adequately resolve the viscous boundary layer. The rest of the flow domain is very large for a typical turbine and thus continuing with the small elements away from the blade would produce a mesh that would be very computationally expensive to solve. Several techniques for reducing the number of elements and thus the mesh size must be employed to obtain an efficient and effective computation.

One such technique is to reduce the overall flow domain to only the portion that is taken up by a single blade rather than all three. This can be accomplished by employing symmetric boundary conditions about the axis of rotation.

'As can be seen the domain is quite large in comparison to the blade and hub. This is because the wake produced by the turbine extends far downstream and it is not desirable for it to extend into the rear boundary'. The flow volume depicted in Fig. 4 is for a wind turbine blade '20 m in length and extends 550 m downstream and 100 m upstream of the blade. These dimensions are based on previous

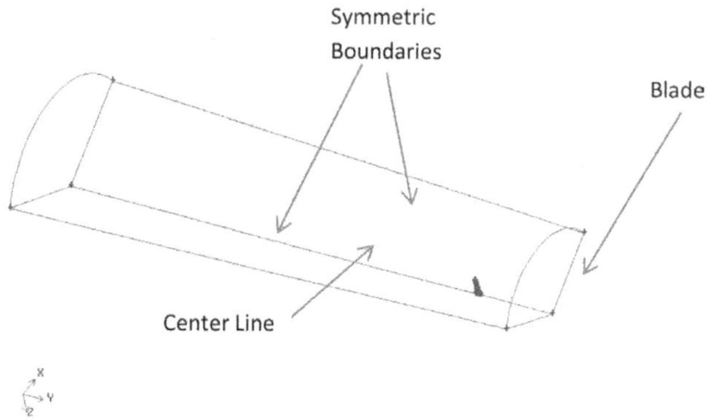

Figure 4: Flow domain for wind turbine CFD simulation.

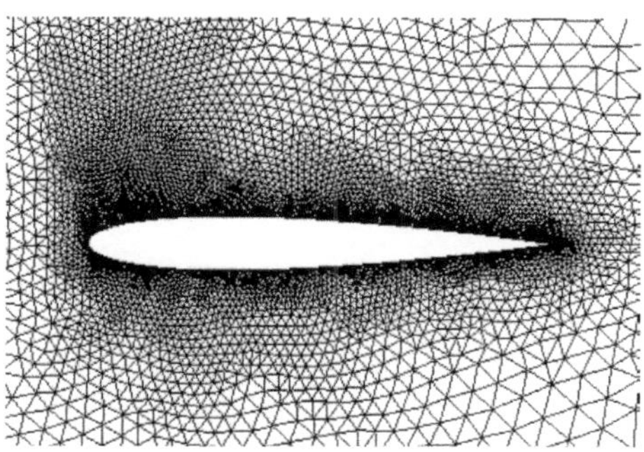

Figure 5: Unstructured mesh.

work [6,7]. The radius of the front face is 100 m while the back is 130 m. The angle between the two side surfaces is 120°. This means that exactly one-third of the wind turbine will be modelled. This is done for computational efficiency'. By specifying symmetric boundary conditions, it is insured that the flow conditions on both planes will be mirrored across the centre line of the flow domain. In most cases, this is a valid assumption because the flow pattern is periodic in the angular direction.

The other technique that is utilized in almost every CFD study is varying the mesh density. In this approach, the mesh elements will be small in areas of interest where high resolution of the flow pattern is required and large in other areas where information about flow conditions is less important or does not change significantly. This will yield less elements than if a small element size was used throughout and more accurate results than if large elements were used throughout.

For a typical wind turbine analysis, 'the area of interest is the region surrounding the aerodynamically active portion of the blade. This portion extends from 2 m above hub to the tip of the blade'. A common meshing technique employed uses both a structured and unstructured mesh. A structured mesh consists ofsix-sided bricks or squares that are stacked in a pattern. An unstructured mesh consists of pyramids or tetrahedral shapes that are assembled in no particular order. An example of each being used to define the flow domain are depicted below in Figs. 5 and 6.

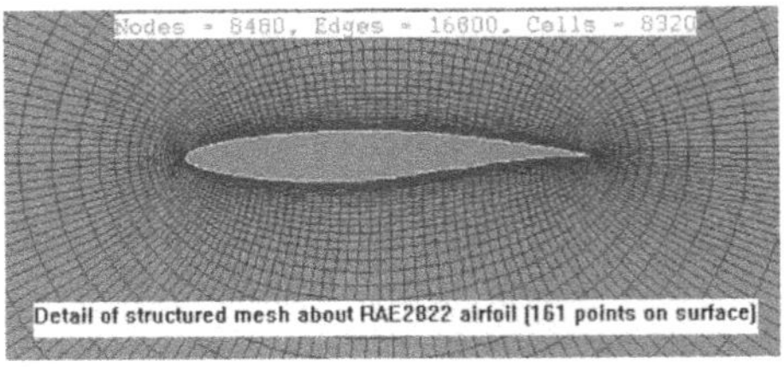

Figure 6: Structured mesh.

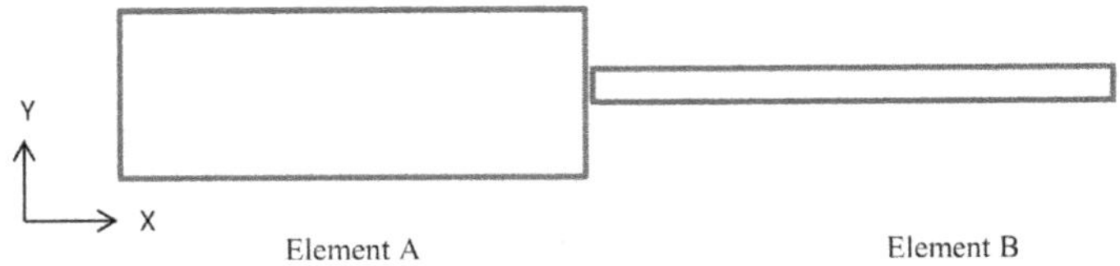

Figure 7: 3D quadrilateral elements.

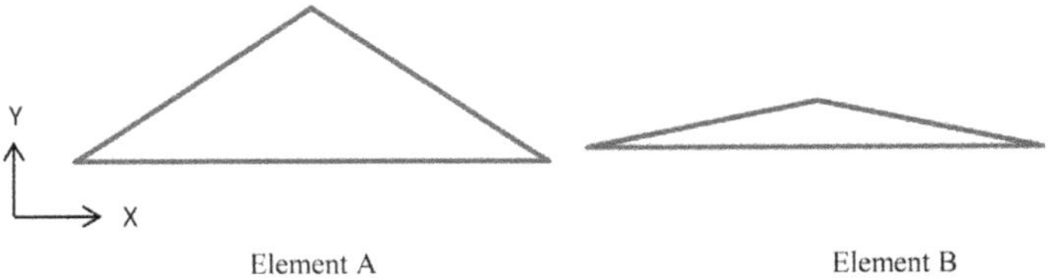

Figure 8: 3D tetrahedral elements.

A structured mesh is recommended in the area surrounding the blade while an unstructured mesh is recommended to resolve the rest of the flow domain. A structured mesh yields more accurate results although it is cumbersome to create while an unstructured mesh is just the opposite. Structured mesh of a 3-D flow domain has certain advantages over unstructured mesh. In particular, the aspect ratio of the elements of a structured mesh can easily be altered to accomplish resolution in a given direction while maintaining their other dimensions and thus integrity.

In Fig. 7, a cross-section of two 3-D quadrilateral elements is depicted. Both elements have the same length of dimension into the page. Element A can be adjusted to element B if property flux in the Y direction contains higher gradients and thus requires higher resolution. The integrity of element B is maintained and calculation of flux through it is straightforward. This is not true if the same transformation is attempted with a tetrahedral as shown in Fig. 8.

Here element B is highly skewed and will hinder the computation. The only way to accomplish high resolution in the Y direction is to shorten the base of the element and increase the overall number of elements. It is for this reason that quadrilaterals are preferred to resolve the flow region in the near wall region where resolution in the direction perpendicular to the blade surface is required.

Since the only area of interest is the area surrounding the blade, a structured mesh can be used here. There are many strategies to constructing structured

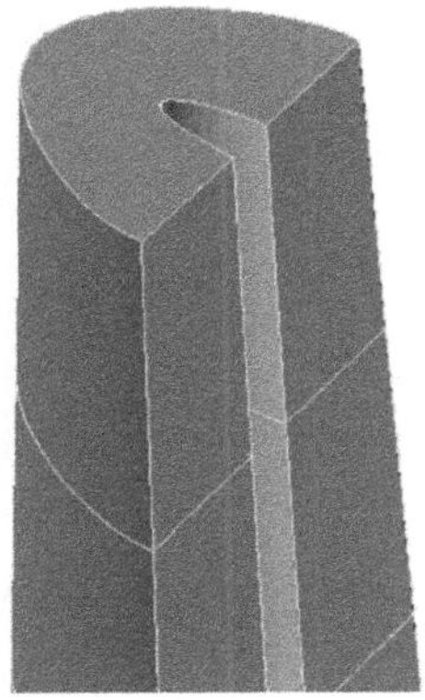

Figure 9: Frontal volume.

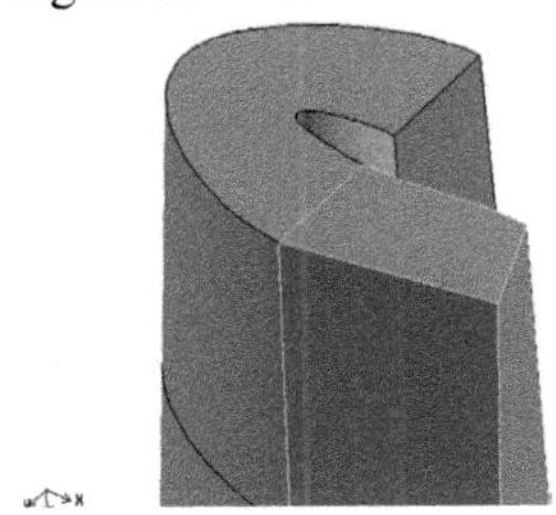

Figure 10: Left side volume.

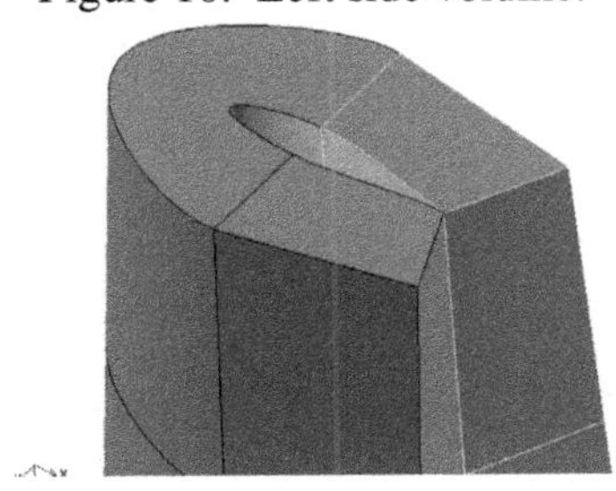

Figure 11: Right side volume.

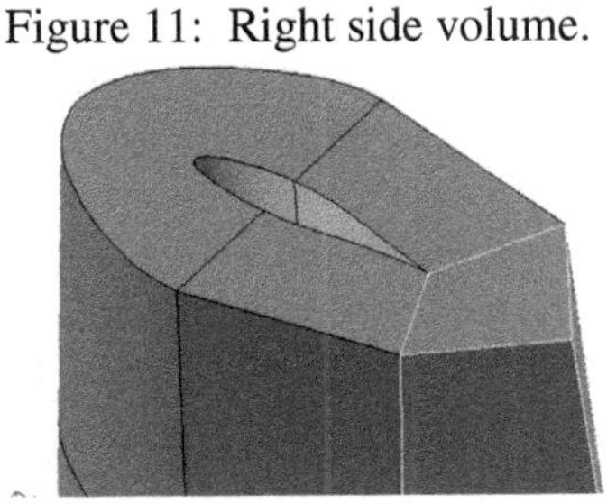

Figure 12: Rear volume.

meshes. Nearly all involve creating subvolumes within the flow domain that can easily mapped or divided up into rectangles to accommodate a structured mesh. One strategy is to create four volumes in the area surrounding the blade as shown in Figs. 9–12.

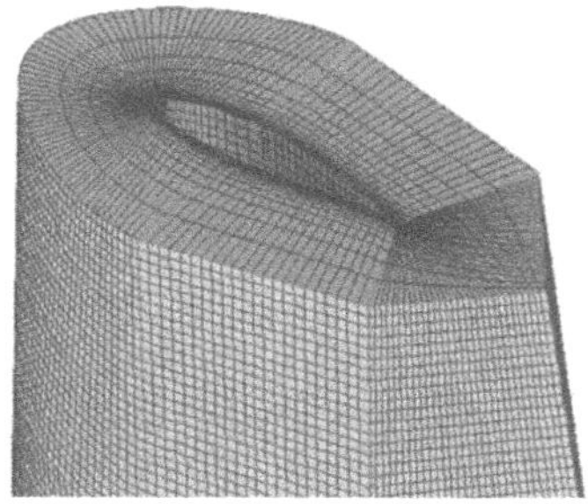

Figure 13: Mesh scheme for volumes surrounding blade.

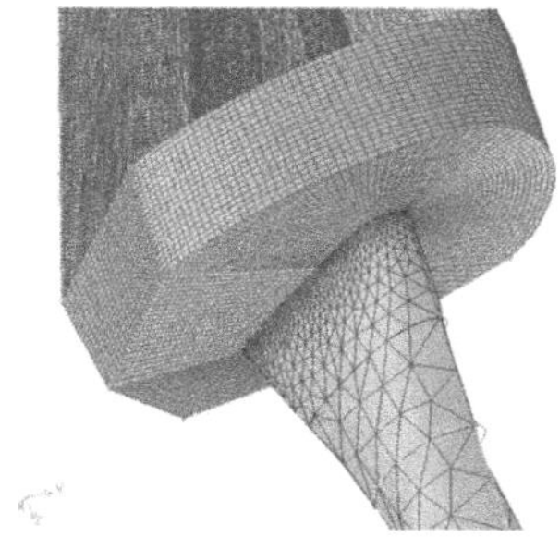

Figure 14: Mesh at bottom of aerodynamically active portion of blade.

Table 1: Mesh spacing near the blade surface.

Starting distance from blade surface (m)	Growth rate	No. of rows
0.00005	1.3	17

When all four volumes are meshed using a 'MAP' scheme typically found in most meshing programs, cubic volumes are produced. This is generally done by meshing opposite faces of each volume first ensuring that the resulting meshes are the same in terms of number of elements. The resulting mesh is shown in Fig. 13.

As can be seen the mesh is considerably denser near the blade surface. This area is of high importance because of the sharp change in air velocity that occurs between the blade surface, where it is zero, and the edge of the viscous boundary layer where it is close to the free stream velocity. Enough elements need to be within this space to adequately capture the flow dynamics that occur in the near wall region, which are ultimately responsible for the overall flow conditions across the blade. An important parameter in determining the quality in this area of the mesh is the y^+ which is the non-dimensional distance from a wall as defined in eqn (1)

$$y^+ = \frac{y\sqrt{\tau_w/\rho}}{\nu}. \tag{1}$$

In eqn (1), y is the distance from the wall, τ w is the shear stress near the wall, ρ is the density of the air and υ is the kinematic viscosity. The first node from the wall should be within a distance of $y^+ = 10$. The parameters listed in Table 1 typically will achieve $0.3 \le y^+ \le 5$ with an average of about 2, which is the adequate boundary layer resolution as shown by other researchers [6].

The cubes are also more closely packed at the leading edge of the blade. This is because this is the location of the stagnation point and there are high-pressure gradients present. It is necessary to adequately resolve these gradients so that the flow over the rest of air foil is not miscalculated.

The four volumes can be extended to the bottom of the aerodynamically active portion of the blade. At the bottom of the volumes, the elements should be evenly spaced so that the tetrahedral unstructured mesh will attach without highly skewed elements as shown in Fig. 14.

The four volumes can also be extended upwards past the tip of the blade in the other direction. A fifth volume then needs to be created in the space between them with the bottom surface of the volume being the top surface of the blade and the top surface being even with the four volumes as shown in Fig. 15.

The four volumes surrounding the centre void are meshed with the MAP scheme while ensuring that the nodes on the outer faces are evenly spaced to prevent skewed elements when an unstructured tetrahedral mesh is attached as shown in Fig. 16.

As mentioned before, the rest of the flow volume can be meshed with a tetrahedral unstructured mesh. It is recommended that the density be varied with the densest mesh at the outside surfaces of the volumes surrounding the aerodynamically active portion of the blade. This can be accomplished with the use of an element growth function found in most meshing software with the parameters stated in Table 2.

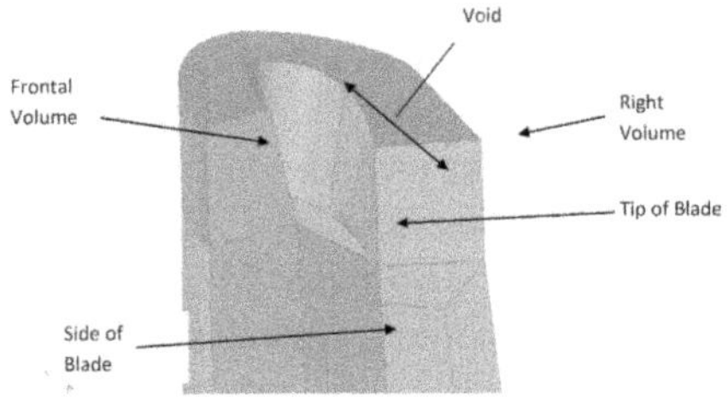

Figure 15: Volumes created at tip of blade.

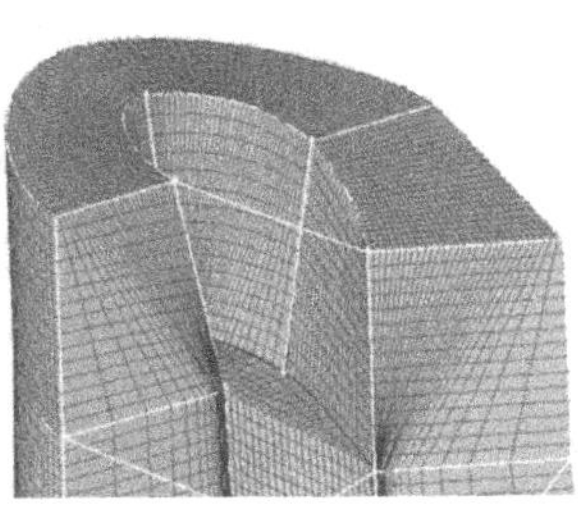

Figure 16: Mesh at tip of blade.

Table 2: Size function parameters used to mesh outer flow volume.

Starting size	Growth rate	Max size
0.03	1.15	15

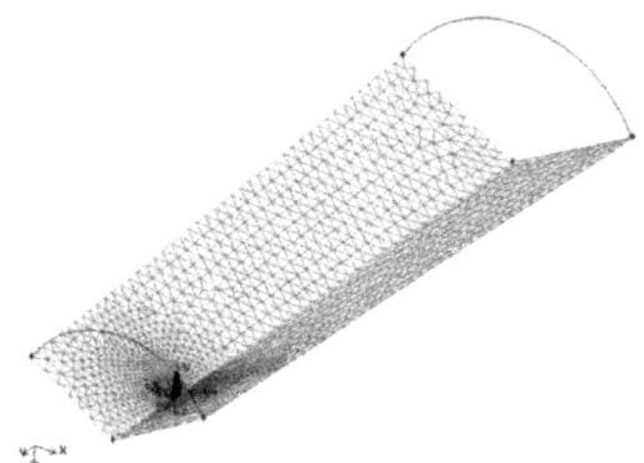

Figure 17: Flow volume mesh density.

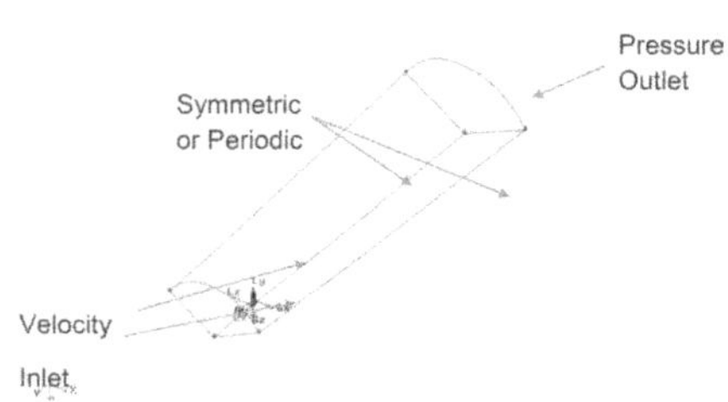

Figure 18: Boundary conditions.

Before this mesh was created the two long rectangular sides were linked to be meshed. This is necessary in order to employ periodic boundary conditions as mentioned earlier. An example of the mesh density is shown in Fig. 17.

Once the flow volume has been meshed it can be exported in a format that can be solved by a CFD solver. Boundary conditions should then be specified along with the appropriate mathematical models to be solved.

Typical boundary conditions are depicted in Fig. 18.

The front and top surfaces should be specified as velocity inlets and in component form. A velocity magnitude should be specified only in the axial direction. The rear boundary should be specified as a pressure outlet with a zero gauge pressure. The hub and blade surfaces should be specified as wall with a no slip condition. To accomplish the rotation rate of the blade, the entire flow volume should be specified as a moving reference frame with a rotation rate corresponding to desired design parameters. This function supplements the equations of motion with an angular acceleration term as follows:

$$\vec{v}_r = \vec{v} - \vec{u}_r, \tag{2}$$

where

$$\vec{u}_r = \vec{\omega} \times \vec{r}. \tag{3}$$

In eqn (2) $\vec{v}_r$ represents a modified velocity vector, $\vec{v}$ represents the velocity vector associated with a stationary reference frame and $\vec{u}_r$ represents the velocity associated with the rotation which is found by multiplying the angular velocity, $\vec{\omega}$, by the distance from the axis of rotation, $\vec{r}$. In effect the flow volume is rotating while the blade is stationary instead of the other way around. This is made possible with the implementation of periodic boundary conditions. This turbulence model and moving reference frame function have been used with success in similar applications with good results [3,7].

4 Numerical and Mathematical Models

Most commercial CFD codes will be able to solve the governing fluid flow and turbulence equations with very little input from the user. Only a brief overview of the equations will be covered in this text.

From a momentum balance analysis of a control volume, the momentum equation is derived and is stated in eqn (4) in tensor notation

$$\frac{\partial U_i}{\partial t}+\frac{\partial U_i U_j}{\partial x_j}=-\frac{1}{\rho}\frac{\partial P}{\partial x_i}+\frac{\partial}{\partial x_j}\left[\nu\left(\frac{\partial U_i}{\partial x_j}+\frac{\partial U_j}{\partial x_i}\right)-\left\langle u_i u_j\right\rangle\right], \tag{4}$$

where U is the average velocity, P is the pressure, ρ is the density, υ is the kinematic viscosity and u is the instantaneous velocity. For this analysis, we are assuming steady-state conditions so the first term on the left is neglected. The air density is assumed constant since the Mach number from the fastest moving part of the blade, the tip, is generally well below 1 (≈.17), which is the level at which compressibility effects are noticeable. From a mass conservation analysis of the same volume, a continuity equation is yielded and is stated in eqn (8) in tensor notation

$$\frac{\partial U_i}{\partial x_i}=0. \tag{5}$$

The turbulence closure model that is most widely used is the k–ω SST (shear stress transport) [4]. There are several differences between the standard and the SST models; the main one being that the *k*–ω SST is a combination of the *k*–ω standard model and the *k*–ε model. This is done because the *k*–ω standard model is dependable and accurate in the near wall region and the *k*–ε is the same but in the farfield region. A blending function is used to combine them is such a way that in near the wall region the *k*–ω model is activated and at greater distances the function goes to zero and the *k*–ε model is used. The *k*–ω SST model is a two-equation model and both parts are expressed in eqns (6) and (7)

$$\frac{\partial}{\partial t}(\rho k)+\frac{\partial}{\partial x_i}(\rho k u_i)=\frac{\partial}{\partial x_j}\left(\Gamma_k\frac{\partial k}{\partial x_j}\right)+\tilde{G}_k-Y_k+S_k, \tag{6}$$

$$\frac{\partial}{\partial t}(\rho\omega)+\frac{\partial}{\partial x_i}(\rho\omega u_i)=\frac{\partial}{\partial x_j}\left(\Gamma_\omega\frac{\partial\omega}{\partial x_j}\right)+G_\omega-Y_\omega+D_\omega+S_\omega. \tag{7}$$

The $\tilde{G}_k$ term represents the generation of kinetic energy, k, from mean velocity gradients and is expressed in eqn (8)

$$\tilde{G}_k=\min(G_k,10\rho\beta^*\kappa\omega), \tag{8}$$

where

$$G_k=-\rho\overline{u_i'u_j'}\frac{\partial u_j}{\partial x_i}. \tag{9}$$

G_ω term represents the generation of ω and is defined in eqn (7). Note that this formulation is different from the standard k–ω model as well as the definition of α∞

$$G_w=\frac{a}{v_t}G_k, \tag{10}$$

where

$$a=\frac{a_\infty}{a^*}\left(\frac{a_0+\mathrm{Re}_t/R_\omega}{1+\mathrm{Re}_t/R_\omega}\right), \tag{11}$$

where

$$a_\infty=F_1a_{\infty,1}+(1-F_1)a_{\infty,2}, \tag{12}$$

$$a^*=a_\infty^*\left(\frac{a_\infty^*+\mathrm{Re}_t/R_k}{1+\mathrm{Re}_t/\mathrm{Re}_k}\right) \tag{13}$$

and

$$a_{\infty,1}=\frac{\beta_{i,1}}{\beta_\infty^*}-\frac{\kappa^2}{\sigma_{\omega,1}\sqrt{\beta_\infty^*}}, \tag{14}$$

$$a_{\infty,2}=\frac{\beta_{i,2}}{\beta_\infty^*}-\frac{\kappa^2}{\sigma_{\omega,2}\sqrt{\beta_\infty^*}}, \tag{15}$$

and where

$$\mathrm{Re}_t=\frac{\rho\kappa}{\mu\omega}, \tag{16}$$

$$R_k=6, \tag{17}$$

$$a_0^*=\frac{\beta_i}{3}, \tag{18}$$

$$\beta_i = 0.072. \tag{19}$$

Γ_k and Γ_ω stand for the diffusivity of k and ω and are defined in eqns (20) and (21)

$$\Gamma_K = \mu + \frac{\mu_t}{\sigma_k}, \tag{20}$$

$$\Gamma_\omega = \mu + \frac{\mu_t}{\sigma_\omega}. \tag{21}$$

In eqns (20) and (21), σk and σω are the turbulent Prandtl numbers, and μt is the turbulent viscosity as defined in eqns (22)–(23)

$$\mu_t = \frac{\rho k}{\omega} \frac{1}{\max\left[\frac{1}{a^*}, \frac{SF_2}{a_1 \omega}\right]}, \tag{22}$$

$$\sigma_K = \frac{1}{F_1 / \sigma_{K,1} + (1 - F_1) / \sigma_{K,2}}, \tag{23}$$

$$\sigma_\omega = \frac{1}{F_1 / \sigma_{K,1} + (1 - F_2) / \sigma_{\omega,2}}, \tag{24}$$

where

$$F_1 = \tanh(\varphi_1^4), \tag{25}$$

$$\varphi_1 = \min\left[\max\left(\frac{\sqrt{\kappa}}{0.09\omega y}, \frac{500\mu}{\rho y^2 \omega}\right), \frac{4\rho\kappa}{\sigma_{\omega,2} D_\omega^+ y^2}\right], \tag{26}$$

$$D_\omega^+ = \max\left[2\rho \frac{1}{\sigma_{\omega,2}} \frac{1}{\omega} \frac{\partial \kappa}{\partial x_j} \frac{\partial \omega}{\partial x_j}, 10^{-10}\right], \tag{27}$$

$$F_2 = \tanh(\varphi_2^2), \tag{28}$$

$$\varphi = \max\left[2 \frac{\sqrt{\kappa}}{0.09\omega y}, \frac{500\mu}{\rho y^2 \omega}\right]. \tag{29}$$

In the above equations, y is the distance to the next surface. *Yk* represents the dissipation of k by turbulence and is defined in a similar way as it is in the standard model except $f\beta^*$ is a constant equal to 1 where in the standard model it is a function

$$Y_k = \rho \beta^* k \omega, \tag{30}$$

where

$$\beta^* = \beta_i^* \left[1 + \zeta^* F(M_t)\right], \tag{31}$$

$$\beta_i^* = \beta_\infty^* \left(\frac{(4/15) + (\mathrm{Re}_t / R_\beta)^4}{1 + (\mathrm{Re}_t / R_\beta)^4} \right), \tag{32}$$

$$\zeta^* = 1.5, \tag{33}$$

$$R_\beta = 8, \tag{34}$$

$$\beta_\infty^* = 0.09. \tag{35}$$

$Y\omega$ represents the dissipation of ω and has several differences from the standard *k*–ω model as well. In the SST model, the $f\beta$ term is a constant equal to 1 instead of being an equation, whereas the βi term is not a constant but an equation as defined in eqn (34)

$$Y_\omega = \rho \beta_i \omega^2, \tag{36}$$

where

$$\beta_i = F_1 \beta_{i,1} + (1 - F_1) \beta_{i,2}. \tag{37}$$

$D\omega$ is the cross-diffusion term, which is a result of the redefining of the *k*–ε model in terms of *k* and ω, which is necessary to blend the two models as mentioned earlier. The cross diffusion term is defined in eqn (38), whereas its positive counterpart is defined in eqn (27)

$$D_\omega = 2(1 - F_1) \rho \sigma_{\omega,2} \frac{1}{\omega} \frac{\partial \kappa}{\partial x_j} \frac{\partial \omega}{\partial x_j}. \tag{38}$$

In the above equations, the following constants were used: $\sigma k_{,1} = 1.176$, $\sigma_{\omega,1} = 2.0$, $\sigma k_{,2} = 1.0$, $\sigma_{\omega,2} = 1.168$, $\alpha_1 = 0.31$, $\beta i_{,1} = 0.075$ and $\beta i_{,2} = 0.0828$. The other constants in the above equations have the same value as they do in the standard κ–ω model. The *Sk* and S_ω are user-defined source terms, which were set to zero for this investigation. This model is solved with the continuity equation as well as the momentum equations.

The governing equations for viscous, incompressible flow can be described by the Navier–Stokes equations as stated in eqn (1). Due to their mixed elliptic–parabolic behaviour, a pressure correction technique is widely used to solve them. This technique is employed in the SIMPLE algorithm, which couples the pressure and velocity. This method was used in this investigation along with the first-order upwind discretization scheme for all equations. All solutions were run till they were fully converged and all converged below an order of 1×10^4.

5 Aerodynamic Noise Model

To simulate aerodynamically generated noise, the boundary layer noise source model was used. This model is appropriate because the noise generated from a wind turbine does not have distinct tones. Since the sound energy is continuously distributed over a range of frequencies, time consideration is not required. There are some assumptions made in deriving this model that are acceptable for modelling noise generated by a wind turbine. These include high Reynolds number, low Mach number and no mean flow. This model in particularly useful in approximating the contribution to total acoustic energy made by a solid surface as opposed to free flow such as a jet. The model consists of Curle's integral [8] based on the Lighthill acoustic analogy [9] as stated in eqn (39)

$$p'(\vec{x},t) = \frac{1}{4\pi a_o} \int \frac{(x_i - y_i)n_i}{r^2} \frac{\partial p}{\partial t}(\vec{y},\tau)\, \mathrm{d}S(\vec{y}). \tag{39}$$

In the above equation, *ao* is the speed of sound, *t* is the time, *x* and *y* are position coordinates, *S* is the integration surface and τ is the emission time and is defined in eqn (40)

$$\tau = t - r / a_o. \tag{40}$$

Using this equation, the sound intensity at distances away from the surface can be found using eqn (41)

$$\overline{p'^2} \approx \frac{1}{16\pi^2 a_o^2} \int_s \frac{\cos^2\theta}{r^2} \overline{\left[\frac{\partial p}{\partial t}(\vec{y},\tau)\right]^2} A_c(\vec{y})\, \mathrm{d}S(\vec{y}), \tag{41}$$

where

$$r = |\vec{x} - \vec{y}|. \tag{42}$$

In eqn (41) A_c is the correlation area, and θ is the angle between *r* and the direction normal to the wall, *n*. Total acoustic power can then be found by integrating eqn (41) over the surface of the body yielding eqn (43)

$$P_a = \frac{1}{\rho_o a_o} \int_0^{2\pi} \int_0^{\pi} \overline{p'^2} r^2 \sin\theta \, \mathrm{d}\theta \, \mathrm{d}\psi \tag{43}$$

6 Moving Reference Frame Model

Along with the momentum, continuity and turbulence equations, a typical moving reference frame function can be utilized. This function supplements the equations of motion with an angular acceleration term as follows:

$$\vec{v}_r = \vec{v} - \vec{u}_r, \tag{44}$$

where

$$\vec{u}_r = \vec{\omega} \times \vec{r}. \tag{45}$$

In eqn (44), $\vec{v}_r$ represents a modified velocity vector, $\vec{v}$ represents the velocity vector associated with a stationary reference frame and $\vec{u}_r$ represents the velocity associated with the rotation, which is found by multiplying the angular velocity, $\vec{\omega}$, by the distance from the axis of rotation, $\vec{r}$. In effect, the flow volume is rotating while the blade is stationary instead the other way around. This is made possible with the implementation of periodic boundary conditions. This turbulence model and moving reference frame function have been used with success in similar applications with good results [3,7].

7 Results Analysis

Once the simulation is set up, it can be run and the data can be analysed. To insure that the simulation was setup correctly, it is common practice to simulate a condition that has been experimentally measured so the results can be compared. A typical wind turbine blade that is used as a simulation benchmark is the LM19.1 blade whose performance was measured on a NTK500/41 turbine. A plot of experimental and simulation performance data is shown in Fig. 19 [10].

Once the simulation parameters have been verified through comparison to experimental results, they can be used to evaluate other geometries under similar conditions. Once the simulations are complete there is a generous amount of both qualitative and quantitative data to evaluate.

If the same parameters are used in the simulation referenced in Fig. 19, it is used to evaluate a straight and a swept edge turbine blade pressure contours as well as a similar power curve can be generated and evaluated as shown in Fig. 20.

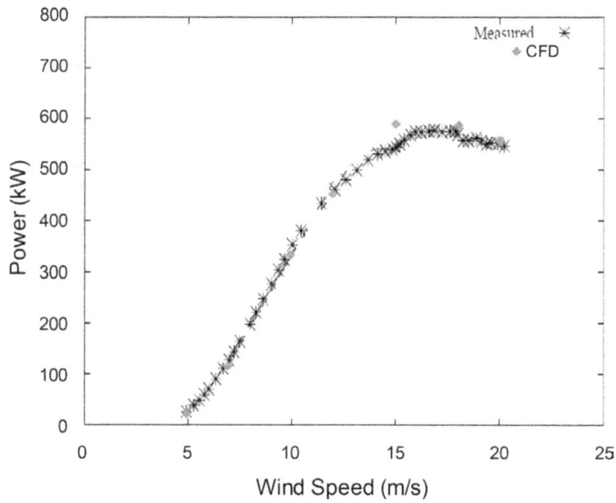

Figure 19: Power generation as a function of wind speed [10].

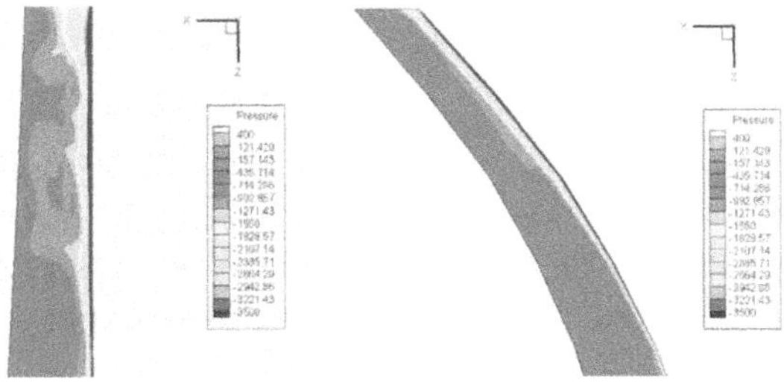

Figure 20: Pressure contour on the blades with straight (left) and swept (right).

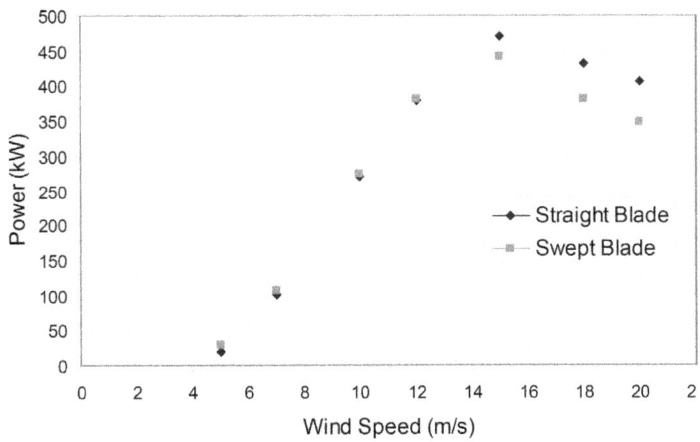

Figure 21: Power curve for both straight and swept blades.

The qualitative data from the pressure contour plots help explain the quantitative data plotted in the power curve (Fig. 21). As can be observed at 20 m/s, the straight edge blade outperforms the swept edge blade in terms of power output. From the pressure contour plot, it is observed that flow separation has not fully occurred on the straight edge blade leaving large portions at lower pressures than are present on the swept edge blade. This information can be used to further alter the design in an attempt to further optimize the design.

8 Conclusions

Wind turbine blade analysis is an integral part of producing a cost-effective wind turbine. Due to their size and cost, evaluation of their performance has largely been done through computer simulation. This process has evolved over the years and is becoming even more sophisticated and consequently invaluable. With the use of the k–ω SST turbulence model along with a mesh that accomplishes an adequate y^+ at the blade surface, which utilizes a moving reference frame environment, turbine blade performance can be accurately predicted in terms of power output while the flow conditions can be studied in great detail.

References

[1] Piggott, H., Small wind turbine design notes. http://worldtracker.org/media/library/Free%20Energy/Free%20Energy/Wind%20Energy%20-%20Small%20Wind%20Turbine%20Design%20Notes.pdf, 2007.

[2] Larwood, S., Zuteck, M,. Davis, U.C. & Consulting, M.D.Z., Swept wind turbine blade aeroelastic modeling for loads and dynamic behavior, AWEA *Wind Power,* **117**, 2006.

[3] Ferre, E. & Munduate, X., Wind turbine blade tip comparison using CFD. *Journal of Physics: Conference Series*, **75**, p. 012005, 2005.

[4] Menter, F.R., Zonal two equation k–ω turbulence models for aerodynamic flows. AIAA-Paper-932906, 1993.

[5] Rasmussen, F., Blade and rotor loads for Vestas 15. Riso-M-2402, Riso National Laboratory, Roskilde, Denmark, 1983.

[6] Bak, C., Fuglsang, P., Sorensen, N., Madsen, H., Shen, W. & Sorensen, J., Airfoil characteristics for wind turbines, Riso-R-1065(EN), 1999.

[7] Mandas, N., Cambuli, F. & Carangiu, C.E., Numerical prediction of horizontal axis wind turbine flow. *European Wind Energy Conference Proceedings*, 2006.

[8] Curle, N., The influence of solid boundaries upon aerodynamic sound. *Proceedings of the Royal Society of London, Series A, Mathematical and Physical Sciences*, **231**, pp. 505–514, 1955.

[9] Lighthill, M.J., On sound generated aerodynamically. I. General theory. *Proceedings of the Royal Society of London. Series A, Mathematical and Physical Sciences*, **222,** pp. 1–34, 1954.

[10] Amano, R.S., Avdeev, I., Malloy, R.J. & Shams, M.Z., Power, structural, and noise performance tests on a different wind turbine rotor blade design. *International Journal of Sustainable Energy*, **32(2)**, pp. 78–95, 2013.

CHAPTER 6

On Ice Accretion for Wind Turbines and Influence of Some Parameters

J.G. Pallarol, B. Sunden & Zan Wu
Department of Energy Sciences, Lund University, Sweden.

Abstract

The aim of this chapter is to find out the influence of the ice accretion on the wind turbine blade performance, to reveal how different icing variables affect ice accretion and to estimate the heat amount required to de-ice the blades at different icing events. Also different computational model simulations of ice accretion and the different de-icing technologies existing today are described. All the simulations of icing events are performed with a software called LEWINT, developed by NASA Glenn Research Centre, USA. The study emphasises the conditions that are most severe for ice accretion. However, the chapter is theoretical, so experiments are needed to validate and complement the conclusions reached in this chapter. It is pointed out that it is important to find the conditions for which ice accretion is formed and the heat amount required to avoid it, to design suitable blades for regions where ice accretion is formed regularly and develop suitable de-icing systems

Keywords: Anti-icing and de-icing, cold climate, ice accretion, ice thickness, wind turbine.

1 Introduction

Wind turbines in cold climates are exposed to icing conditions and low temperatures outside the design limits of standard wind turbines. Standard turbines operating in such extreme environments are prone to production losses and increased loads, which in turn will cause a risk of premature mechanical failure and financial losses. Although exact numbers are hard to assess, some 60 GW of installed capacity is located in cold climate (CC) areas in Scandinavia, North America, Europe and Asia. Narrowing this assessment only to regions with high likelihood of extended CC operation; Scandinavia, Switzerland, Canada and northern parts of the United States and China, wind turbines to the power of 20 GW were conservatively estimated to be installed in CC areas in 2011. Additionally, microclimates with similar conditions are found in more temperated areas such as central and southern Europe, China, Japan, many parts of the United States and locations in the southern hemisphere such as Australia, New Zealand and southern South America. Recent interest in offshore wind power development introduces further demands for knowledge in dealing with icing, as turbines installed in the shallow waters in northern Europe and in the coast of New England in the United States also face icing conditions.

Sites within CC constitute a vast wind energy production potential. Because fewer temperate sites are available and the offshore wind development faces higher costs than expected, large wind energy projects in CC become tempting and are likely to be implemented. Increased experience and knowledge, combined with improvements in technology directed to be used within CC, have enabled such projects to become more competitive when compared with those onshore with low wind resources and those offshore to be built at higher costs.

1.1 Wind energy

The power available in the wind for a wind turbine for use is given by

$$Pw = 0.5\rho_a A v_w^3 \quad (1)$$

where ρ_a is in kg/m^3, A is in m^2 and υ_w is in m/s, so Pw is in Watts. This means that the mechanical power depends mostly on the wind speed. A doubling of the wind speed will increase the power eight times. In Fig. 1, an example of the dependence of power with wind speed is shown. The range of wind speed depends on the power capacity of the turbine.

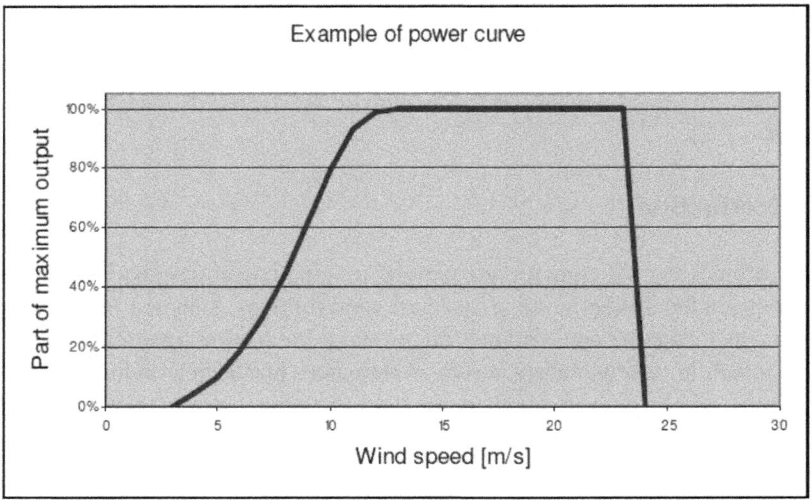

Figure 1: Dependence of power on wind velocity.

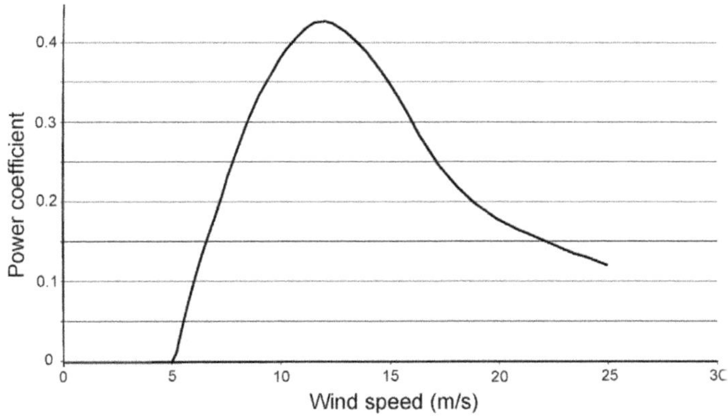

Figure 2: Dependence of *Cp* on wind velocity [1].

After having described the power available, a presentation on how the wind turbines extract energy from the wind is given. Wind turbines extract energy by slowing down the wind, the wind hits the turbine blades and the energy sets the blades in rotation. Air blowing into a wind power station has to pass through the rotors but must have energy left for blowing away. Due to this, the maximum fraction of the energy which can be extracted is 0.59. This is known as the Betz limit. However, a wind turbine normally has an efficiency lower than this value. The constant used for the efficiency is called *Cp*, so the power produced by a wind turbine is

$$P = PwCp \tag{2}$$

where *Pw* comes from eqn (1). Modern wind turbine blades are designed from one or multiple aerofoils, resulting in power coefficients of almost 0.5. The turbine efficiency *Cp* is a function of the tip speed ratio, for a given wind speed. This ratio is defined as

$$\lambda_{sr} = \frac{V_{tip}}{V_w} = \frac{w_r R}{V_w} \tag{3}$$

where V_{tip} is the speed of the blade tip, V_w is the wind speed, w_r is the rotational speed of the rotor and R is the radius of the rotor.

It is shown in Fig. 2 how the power efficiency coefficient evolves with the wind speed, but every turbine has its own *Cp* curve. In this case, the maximum value of the coefficient is achieved when the wind speed is 12 m/s. However, the effect of wind speed is much higher than the effect of the efficiency coefficient, so the optimal point is not the one at the best *Cp*.

2 Cold Climates

Wind turbines in cold climates refer to sites that may experience significant time or frequency of either icing events or low temperatures outside the operational limits

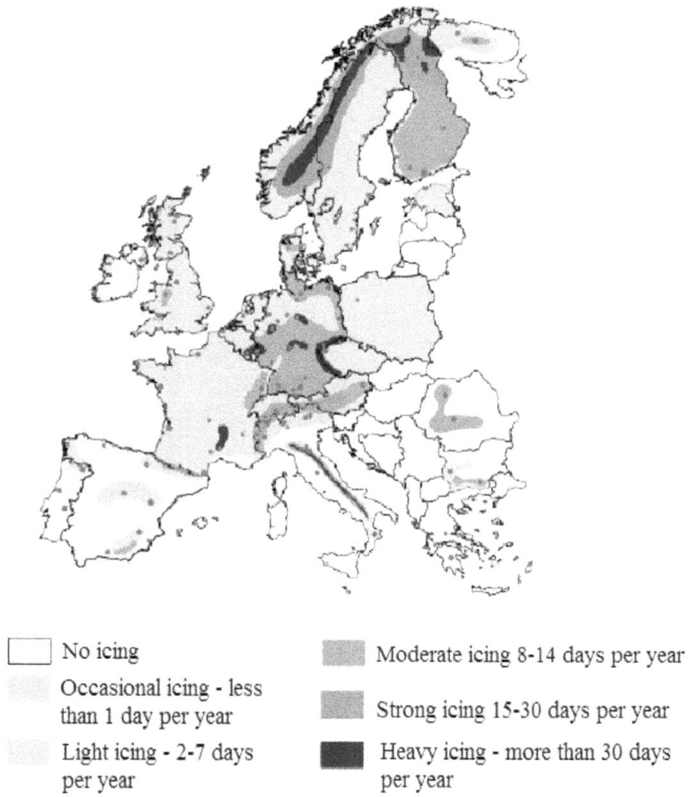

Figure 3: Regions with cold climates [3].

of standard wind turbines. Apart from lower energy production, or even no production, which directly influences a wind farm's economy, there are legal issues, such as ice throw and increased noise, fatigue loading and operations and maintenance aspects that need to be considered [2]. Areas, where periods with temperatures below the operational limits of standard wind turbine occur, are defined as low-temperature climate (LTC) and areas with icing events are defined as icing climate (IC). A region can be in a LTC or in a IC or both while they are still all denoted Cold Climate Sites. In Fig. 3, a description of different CC regions in part of Europe is given.

2.1 Low-temperature climate

Extremely low temperatures are typically caused by clear sky radiation during the winter time, which is often associated with high pressure zones. Furthermore, the elevation of a site plays an important role.

An LTC can have different effects on a wind energy project:

- Materials used in turbines can be affected by low temperatures.

- Maintenance work at low temperatures is more time-consuming.
- Cold start of wind turbine can be more difficult at low temperatures.
- Lubricants can lose their viscosity.
- Internal heat of wind turbines reduces energy production.

2.2 Icing climate

Atmospheric icing is defined as the accretion of ice or snow on structures, which are exposed to the atmosphere. There are two different types of atmospheric icing, in-cloud icing (rime ice or glaze ice) and precipitation icing (freezing rain, wet snow). The different forms of atmospheric icing are:

- *Rime ice*: Super-cooled liquid water droplets from clouds or fog are transported by the wind. When they hit a surface they freeze. Rime icing is common at high altitudes and at low temperatures. The most severe rime icing event occurs on freely exposed mountains or hilltops where air is forced upwards and consequently cooled. Depending on droplet size and air temperature during the icing event, rime ice can form structures of different density and strength, which leads to a division into two types of rime ice, hard and soft rime ice. Hard rime ice is opaque, usually white and it adheres firmly on surfaces making it very difficult to remove. The density of hard rime ice ranges between 600 and 900 kg/m^3 according to ISO 12494. Soft rime ice is thin ice with needles and flakes. The growth of soft ice starts usually at a small point and grows triangularly into the windward direction. The density of it is between 200 and 600 kg/m^3 according to ISO 12494, and it can be more easily removed. In Fig. 4, there is an example how rime ice is formed [4].
- *Glaze ice*: It is caused by freezing rain, freezing drizzle or wet in-cloud icing and forms a smooth, transparent and homogeneous ice layer with a strong adhesion on the structure. Traditionally, glaze ice forms at temperatures between 0°C and –4°C. Part of the water droplets freeze upon impact and remainder run along the surface before freezing, forming a profile shape of high density clear ice. Its density is around 900 kg/m^3 according to ISO 12494. Freezing rain or drizzle occurs when warm air melts snow crystals and forms rain droplets, which afterwards fall through a freezing air layer near the ground. Wet in-cloud icing occurs when the surface temperature is near 0°C, the water droplets which

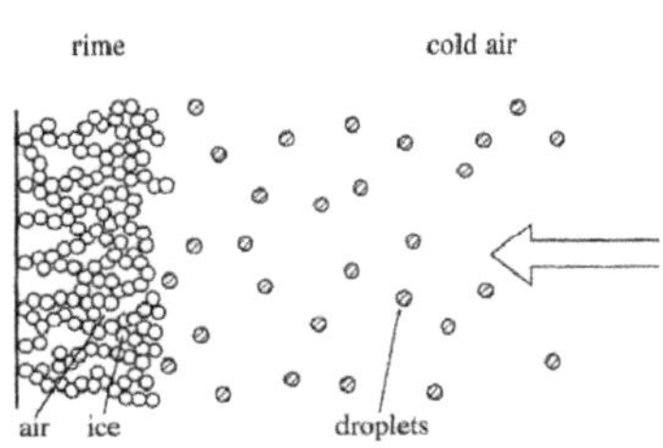

Figure 4: Rime ice [4].

hit the surface do not freeze completely. A layer of liquid water forms, which may flow around the object and freeze also on the leeward side. Figure 5 shows the formation of glaze ice and the difference with rime ice [4].

- *Wet snow*: Partly melted snow crystals with high liquid water content (LWC) become sticky and are able to adhere on the surface of an object. Wet snow accretion, therefore, occurs when the air temperature is between 0°C and 3°C. The typical density is 300–600 kg/m^3. The wet snow will freeze when the wet snow accretion is followed by a temperature decrease [4].
- *Hoarfrost*: It is formed when water vapour sublimes to ice. This happens at very low temperatures and will form ice with very low density, which is normally not considered to cause any significant load on a structure.

2.3 Effects of icing on wind energy

Climatic conditions with turbulence, gusts, icing and lightning strikes in cold regions will affect wind turbine performance and cause idling of wind farms on windy days. Particularly, icing affects the wind assessment and the operation of wind farms. The following problems are related to icing and cold climates.

- *Measurement errors*: During the assessment phase, the anemometers, wind vanes and temperature sensors can be affected by ice. At icing conditions, wind speed errors can be as high as 30%. Another study identifies a maximum error of 40% for an ice-free anemometer and 60% for a standard anemometer during icing events [5].
- *Power losses*: Ice accretion changes the shape and roughness of the blade aerofoil (consequently affecting their aerodynamic characteristics) and introduces measurement errors from turbine instruments (wrong wind speed or direction, which affects yaw and power controls). Small amounts of ice on the leading edge of aerofoils significantly reduce aerodynamic properties of the blade and the resulting power production. Power losses may vary from 0.005% to 50% of the annual production, depending on icing intensity and its duration on the site, wind turbine models and the evaluation methodology [5].
- *Overproduction*: Higher air density related to low temperatures and aerofoil modifications can lead to overproduction of the wind turbine. Overproduction of up to 16% has been recorded.

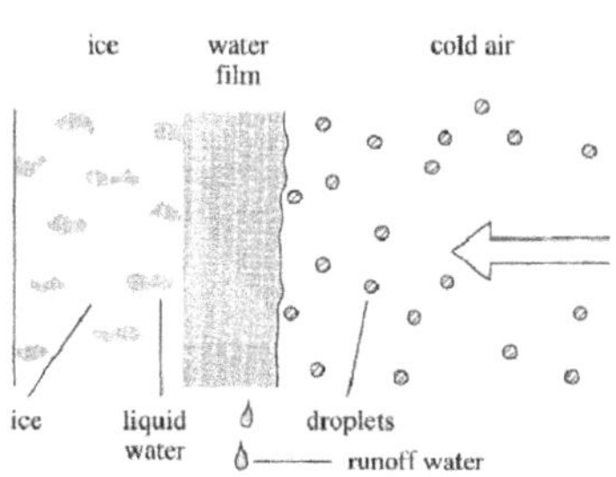

Figure 5: Glaze ice [4].

- *Mechanical failures*: Ice accretion will increase the load on the blades and on the tower structure, causing high amplitude vibrations and/or resonance as well as mass imbalance between blades. Operation at low temperatures affects oil viscosity and changes the dimensions and mechanical properties of different components of the wind turbine. This results in possible overheating and higher fatigue charges on components.
- *Electrical failure*: Snow infiltration in nacelle and extreme temperature lead to condensation in the electronics [5].
- *Safety hazard*: The pieces of ice shed from wind turbines can be quite large, and are definitely not insignificant. It is likely that something will occur eventually if preventive measures are not implemented. Though the risks are greatest for service personnel who must approach the wind turbine, others can be at risk when the wind turbine is located near a road or recreation area [5].

3 Ice Accretion

3.1 Physical model

Although estimation of icing on wind turbines and other structures is very important, no undisputed model has so far been developed. Due to the non-linear relationships between the parameters affecting ice accretion, most models include empirical formulas and none can be solved analytically. The most widely used model today is Makkonen's algorithm for ice accretion. The Makkonen model [6] is originally developed for modelling of ice accretion on overhead power lines, and therefore models ice accretion on a cylinder resembling a power line. Now the Makkonen model will be explained and it is based on reference [6].

Irrespective of what kind of particles, liquid or solid, that cause ice on an object, the accumulation rate will be dependent on the flux density; in other words, the cross-sectional area of the object, the velocity of the particles and the mass concentration of particles in the air. Because not all particles will collide with the object and not all particles that do collide will stick or accrete, three efficiency coefficients need to be introduced

$$\frac{\mathrm{d}M}{\mathrm{d}t} = a_1 a_2 a_3 w v A \tag{4}$$

where a_1 is the collision efficiency, a_2 is the sticking efficiency, a_3 the accretion efficiency, w the mass concentration, υ the particle velocity relative to the object and A the cross-sectional area relative to the particle velocity. The Makkonen equation can easily be used to model ice load and ice accretion rate. It is common to give the ice load and ice accretion rate in kg/m and kg/m/hr, respectively. The latter refers to per metre of the cylinder used in the model. The ice accretion rate is sometimes presented in terms of active ice hours, which refers to hours with ice accretion rate above a pre-defined threshold, commonly 10 g/m/hr.

3.1.1 The collision coefficient, α_1

The collision coefficient, a_1, is the ratio of the flux density of particles hitting the object to the total flux density. a_1 decreases from 1.0 when particles are forced around the object instead of hitting it. Small droplets, large cross-sections and low wind velocity reduce a_1. The droplet trajectory is determined by the aerodynamic drag and the inertia. For a small droplet, the inertia will be small and it will tend to follow the streamlines around the object instead of hitting it. If the inertia and drag forces are known, the collision coefficient can be calculated theoretically by numerical calculation of the droplet trajectories. Because this is computationally expensive, Makkonen suggests the simplification to assume a cylindrical object, and to use an empirical equation for a_1

$$a_1 = A - 0.028 - C(B - 0.0454) \tag{5}$$

where

$$A = 1.066K^{-0.00616}e^{-1.103K^{-0.688}} \tag{6}$$

$$B = 3.641K^{-0.498}e^{-1.497K^{-0.694}} \tag{7}$$

$$C = 0.00637(\phi - 100)^{0.381} \tag{8}$$

where K and f are dimensionless parameters

$$\phi = \frac{Re_\mathrm{d}^2}{K} \tag{9}$$

where Re_d is the Reynolds number based on the free stream velocity and is given by

$$Re_\mathrm{d} = \frac{\rho_\mathrm{a} dv}{\mu} \tag{10}$$

and

$$K = \frac{\rho_\mathrm{w} d^2 v}{9\mu D} \tag{11}$$

where d is the droplet diameter, D the cylinder diameter, ρ_w the water density, μ the dynamic viscosity of air and ρ_a the air density.

Because a_1 depends on the droplet size, one can safely assume $a_1 = 1.0$ for wet snow and freezing rain as long as the structure studied is not extremely large. This means that a_1 normally needs to be calculated only for occasions with in-cloud icing with small droplets.

3.1.2 The sticking coefficient, α_2

The coefficient a_2 represents the efficiency of collection of those particles that hit the object. It means that a_2 is the ratio of the flux density of the particles that stick to the object to the flux density of the particles that hit the object. The sticking efficiency α_2 is reduced from unity when the particles bounce from the surface. The particles here are considered to stick when they are permanently collected, or their residence time on the surface is sufficient to affect the icing rate due to, for example, heat exchange with the surface. For liquid droplets, it can normally be assumed that they do not bounce and hence α_2 equals 1.0. As soon as there is a liquid layer on the surface, combined with favourable conditions, basically all droplets will stick to the surface. For ice and snow particles, the ratio of particles that bounce off the object can be a significant share. For solid particles, such as dry snow, α_2 is close to or equal to 0.

There is presently no detailed theory for the sticking efficiency of wet snow. The available approximation methods of α_2 are empirical equations based on laboratory simulations and some field observations. The best first approximation for α_2 for cylindrical shapes is probably

$$\alpha_2 = \frac{1}{v} \tag{12}$$

where the wind speed, v, is in m/s. When $v < 1.0$ m/s, $\alpha_2 = 1.0$. The air temperature and humidity also affect α_2, but there are presently not enough consistent data to take them into account, despite several experimental studies on this problem. However, as pointed out above, $\alpha_2 > 0$ only when the snow particle surface is wet. Thus, snow does not accrete effectively when the wet-bulb temperature is below 0°C. This criterion is very important because it allows determination of the duration of wet-snow events by climatic weather data.

3.1.3 The accretion coefficient, α_3

The accretion efficiency coefficient, α_3, is the ratio of icing to the flux density of droplets sticking to the object's surface. During dry growth α_3 equals 1.0 because there is no run-off. During wet growth, glaze icing, there is a liquid film on the adhesion surface and α_3 will be less than 1.0. The freezing rate will then be determined by the rate with which latent heat from freezing can be transported away from the surface. A heat balance over the surface of the object can be used to determine the portion that will freeze

$$Q_f + Q_v = Q_c + Q_e + Q_l + Q_s \tag{13}$$

where Q_f is the latent heat released during freezing, Q_v is the frictional heating of air, Q_c is the loss of sensible heat to air, Q_e is the heat loss due to evaporation, Q_l is the heat loss in warming the impinging supercooled water to the freezing temperature and Q_s is the heat loss due to radiation.

The terms of the heat-balance equation can be parametrised using the meteorological and structural variables.

The heat released in the freezing is transferred from the ice-water interface through the liquid water into the air, and consequently there is a negative temperature gradient ahead of the growing ice. This kind of supercooling results in a dendritic growth morphology, and consequently some liquid water is trapped within the spray ice matrix. Because the unfrozen water can be entrapped without releasing any latent heat, the term Q_f in eqn (13) is

$$Q_f = (1-\lambda)a_3 F L_f \tag{14}$$

where λ is the liquid fraction of the accretion, and F is the flux density of water to the surface ($F = a_1 a_2 w v$). Attempts to determine the liquid fraction, λ, have been made both theoretically and experimentally. These studies suggest that λ is rather insensitive to the growth conditions, and that a value of λ around 0.3 is a reasonable first approximation.

Another important consequence of the existence of porous ice is that the heat conduction into the ice does not need to be considered in the heat balance, because the mixture of ice and water is at the uniform temperature of 0°C and therefore involves no heat conduction through it.

The kinetic heating of air, Q_v, is a relatively small term (except at aircraft speeds), but, because it is easily parametrised by

$$Q_v = \frac{h r v_w^2}{2C_{pa}}, \tag{15}$$

it is usually included in the heat balance. Kinetic heating of the droplets is insignificant and is ignored. Here, h is the convective heat-transfer coefficient, r is the recovery factor for viscous heating ($r = 0.79$ for a cylinder), v_w is the wind speed and C_{pa} is the specific heat of air.

The convective heat transfer is

$$Q_c = h(t_s - t_a) \tag{16}$$

where t_s is the temperature of the icing surface ($t_s = 0$°C for pure water), and t_a is the air temperature.

The evaporative heat transfer is parametrised as

$$Q_e = \frac{h\psi L_e(e_s - e_a)}{C_p p} \tag{17}$$

where ψ is the ratio of the molecular weights of dry air and water vapour (ψ = 0.622), L_e is the latent heat of vaporisation, e_s is the saturation water vapour pressure over the accretion surface, e_a is the ambient vapour pressure in the airstream and p is the air pressure. Here, e_s is a constant (6.17 mbar) and e_a is a function of the temperature and relative humidity of the ambient air.

The term Q_l, is caused by the temperature difference between the impinging droplets and the surface of the icing object

$$Q_l = F C_w(t_s - t_d) \tag{18}$$

where C_w is the specific heat of water, and t_d is the temperature of the droplets at impact.

For cloud droplets, $t_d = t_a$ can be assumed because of the small terminal velocity. This assumption must usually be made also for supercooled raindrops, although they are not fully adjusted to the temperature at the surface layer during their fall.

The heat loss due to long-wave radiation may be parametrised as

$$Q_s = \sigma\, a(t_s - t_a) \tag{19}$$

where σ is the Stefan–Boltzmann constant (5.67×10^{-8} W/m^2/K^4), and a is the radiation linearisation constant (8.1×10^7 K^3). This equation only takes into account long-wave radiation and assumes emissivities of unity for both the icing surface and the environment. Short-wave radiation of the sun is neglected because, in practice, atmospheric icing always occurs in cloudy weather. Using the parametrisation of eqns (14–19) in the heat-balance eqn (13) and solving the accretion efficiency result in the following equation:

$$\alpha_3 = \frac{1}{F(1-\lambda)}\left[(h+6a)(t_s - t_a) + \frac{h\psi L_e}{C_{pa}p}(e_s - e_a)\frac{hrv_w^2}{2C_{pa}} + FC_w(t_s - t_d)\right] \tag{20}$$

3.1.4 Limitations of the model

Although Makkonen's model for ice accumulation is widely used, it is limited by the fact that it models ice accretion on a cylinder. To translate the ice load on an ideal cylinder to a complex structure such as a wind turbine is computationally expensive and so far not possible for the general case. To model ice accumulation on a wind turbine, one needs to break down the structure into small elements, still taking into consideration shadowing effects from other elements and ice growing elements together to form a single element.

One of the more practical problems is that the LWC of the air, the cloud droplet density and the median volume droplet size, MVD, in the air are required values. These parameters are not routinely measured nor does there exist a satisfactorily way to measure them. The average droplet diameter has to be approximated, and the average number of droplets in-cloud has to be known. The LWC can be modelled, but the result has a high uncertainty. There is an uncertainty in the model when α_1, the collision efficiency, is small, because α_1 depends on the droplet diameter and errors are introduced in estimating MVD. Also, the effects of droplet trajectory angle on the collision efficiency are poorly known, which means that the effects of wind direction and rotational speed of a wind turbine are uncertain. Another uncertainty regarding α_1 concerns that low velocities are introducing uncertainties in the Reynolds number and flow patterns.

Modelling of ice accretion also requires modelling of the ice density, because the icing rate depends on the dimensions of the iced object. The density of the ice can be modelled using ballistic models, but for rime-ice best-fit equations have been developed in wind-tunnel experiments

$$\rho_{ice} = 0.378 + 0.425(logR) - 0.823(logR)^2 \tag{21}$$

where R is the Macklin parameter, a dimensionless density parameter. It is given by

$$R = -\frac{v_0 d_m}{2t_s} \quad (22)$$

where v_0 is the droplet impact speed, d_m is the droplet median diameter and t_s is the surface temperature. t_s can be approximated as the air temperature. To solve the equation, the impact speed must first be calculated from the heat balance equation. A difficulty with solving the heat balance equation is the convective heat transfer coefficient. Even assuming a simple structure such as a cylinder, to determine the heat transfer coefficient is complicated.

Another limitation of the model is that it takes into account only accumulation of ice. Ice will be removed only due to the sign of the heat balance. Melting is not included, neither is shedding of ice nor sublimation.

3.2 Computational models

Nowadays, there are several ice accretion codes in the international aircraft icing community such as LEWINT (USA), ONERA (France), DRA (UK), CANICE (Canada) and more recently CIRA (Italy) based on the physics of icing. But the main two codes for ice accretion in wind turbines are TURBICE, which is the most specific code for the field of wind power; and LEWINT, which can be adapted for ice accretion in wind turbines. The second one is the code chosen for the practical part of this chapter.

LEWINT integrates the ice accretion code LEWICE (version 3.2.2) with American Kestrel's user interface, icing analysis tools and automated plotting. LEWICE was developed by the icing branch at the NASA Glenn Research Center in Cleveland, Ohio. The LEWICE base is an ice accretion prediction code that applies a time stepping procedure to calculate the shape of an ice accretion. The atmospheric parameters such as temperature, pressure, wind velocity and the meteorological parameters such as LWC, droplet diameter and relative humidity are specified as input information and used to simulate the shape of the ice accretion. The software consists of four major modules:

1. Flowfield calculation.
2. Particle trajectory and collision calculation.
3. Thermodynamic and ice growth calculation.
4. Modification of the current geometry by adding the ice growth to it.

LEWICE applies a time-stepping procedure to grow the ice accretion. The flowfield and droplet collision characteristics are determined for the clean geometry. The particle trajectories and impingement points on the body are calculated from a potential flow solution that is produced by the Douglas Hess-Smith 2-D panel code included in LEWICE. Alternately and if specified, the flow solution can be obtained from a grid generator and grid-based flow solver or read in as a solution file from this flow solver. The ice growth rate on each segment defining the surface is then determined by applying the thermodynamic model. When a time increment is specified, this growth rate can be interpreted as an ice thickness and the body coordinates are adjusted to account for the accreted ice. The ice growth rate on the surface body is calculated from the icing model that was first developed by Messenger. This procedure is repeated, beginning with the calculation of the flow field about the iced geometry, and continued until the desired icing time has been reached.

LEWICE can perform modelling of both dry and wet (glaze) ice growth. In addition to simulating the ice accretion, LEWICE incorporates a thermal anti-icing function. Calculation of the power density to prevent ice formation is performed by LEWICE. Two anti-icing modes are possible: running wet and evaporative.

It can generate not only anti-icing values, but also data about droplet trajectories, collection efficiencies, collision limits, energy and mass balances, ice accretion shape and thickness. Ice accretion shapes for cylinders and several single-element and multi-element aerofoils have been calculated using this software. The calculated results have been compared to experimental ice accretion shapes obtained both in flight and in the Icing Research Tunnel at NASA Glenn Research Center.

Comparison of available experimental data and LEWICE simulation results shows that based on the overall assessment factor, the variation in the experimental data was 4.4% and the LEWICE predicted ice shape, which differed from the experimental average by 12.5% [7].

4 ADIS

Icing mitigation systems result from two main strategies: anti-icing and de-icing systems (ADIS). Anti-icing prevents ice to accrete on the object while de-icing removes the ice layer from the surface. Both strategies can also be divided into two methods: passive and active. Passive methods take advantage of the physical properties of the blade surface to eliminate or prevent ice, while active methods use external systems and require energy supply that is either thermal, chemical or pneumatic. Many of the strategies from the aerospace industry can be transferred to the wind energy sector, although some scaling has to be done to adjust parameters (wind speed, chord length and aerofoil).

Consequently, because most ADIS are based on heating, wind turbines need more power to operate. The extra power will be added to the consumption of the cold climate package. Also, additional maintenance should be planned. In global values, the de-icing consumption translates into a 1%–4% loss of annual energy production, depending on icing severity [8].

An investment of 5% of the cost of a 600 kW turbine has been estimated for the purchase and installation of ADIS. Cost percentage decreases as turbine size increases. Depending on icing severity on the site and the price of electricity, ADIS' payback time will vary from 1 to 18 years. For a site with medium icing severity, with an average of 30 icing days per year, the payback time should be less than 5 years.

Most of the information described in this section is based on the article by Parent and Ilinca [8]. In adition, some information has been provided by the company ENERCON.

4.1 Passive ADIS

4.1.1 Passive anti-icing systems

The passive anti-icing systems are: black paint, special coating and chemicals.

4.1.1.1 Black paint. Black paint allows blade heating during daylight and is used with an ice-phobic coating. When tested in Yukon (Canada), this method showed immediate and noticeable improvement in performance. This method may be sufficient in sites where icing is slight, infrequent and where icing periods are followed by temperatures above 0°C or in areas with high winter solar intensity at lower altitudes. Most of the time, it is not sufficient to prevent icing. The temperature of the blade's surface may affect the properties of glass-fibre reinforced plastics, as they are sensitive to high temperature. However, another study shows that black blades do not overheat in the summer.

4.1.1.2 Special coating. In theory, ice-phobic coatings prevent ice from sticking to the surface because of their anti-adherent property, while super-hydrophobic coatings do not allow water to remain on the surface because of repulsive features. Reducing the shear forces between the ice and the surface will also reduce sensitivity to dirt and bug. Currently, most manufacturers use epoxy or polyester matrix composites reinforced with glass and/or carbon fibres, although polyester and glass fibres remain the material of choice due to their lower cost. Current research is heading towards nanocomposite coatings, polymers reinforced by minute, nanometre-scale particles. These nano-composites create very high contact angles with water. Advantages and disadvantages are listed below.

Advantages:

- Low cost.
- No special lighting protection needed.
- Easy blade maintenance, protects the whole surface.
- Easy to apply.

Disadvantages:

- Several materials tested but no good solution has been found.
- Icing occurred even on coated surfaces regardless of the temperature.
- Material degradation, coating becomes porous.
- A clean and smooth surface is preferable.

4.1.1.3 Chemicals. When applied on a blade surface, chemicals reduce the water freezing point. It is mostly used during aircraft take-off. It is a pollutant and it needs special application and a lot of maintenance. It cannot remain on the surface of the blade for a long period.

4.1.2 Passive de-icing systems

The passive de-icing systems are flexible blades and active pitching.

4.1.2.1 Flexible blades. Flexible blades help to crack the ice and loose its contact with the surface. Blade flexing is known to help shedding the ice but very little information is available on the subject.

4.1.2.2 Active pitching. The active pitching is a semi-active method that uses start/stop cycles to orient iced blades into the sun. The method may work in slight icing environments but it has not been scientifically validated and may damage wind turbines.

4.2 Active ADIS

4.2.1 Active anti-icing systems

The active anti-icing systems are used to prevent the ice accretion on blade surface and are based on resistance heating, air layer or microwave heating.

4.2.1.1 Resistance heating. Heating resistance and warm air can be used in anti-icing mode to prevent icing. The blade temperature should be kept around 0°C to prevent icing. The advantage is that no ice accumulates on the blades. A blade can be kept at –5°C, instead of 0°C, in good condition. In this way, 33% of power can be saved which represents 2.3% of winter production. The inconvenience is that it requires a lot of energy. If it is used to prevent runback at 100°C, it is close to the softening point of some epoxies and resins (although thermosetting plastics that are designed for higher operating temperatures are available). The continuous operating temperature should be less than 50°C with current blade materials [8].

4.2.1.2 Air layer. The air layer consists of an air flow originating inside the blade and pushed through rows of small holes near the blades' leading and trailing edges to generate a layer of clean and, if necessary, heated air, directly around the blade surface. This method would deflect the majority of water droplets in the air and would melt the few droplets that managed to hit the surface but very little information is available for its application.

4.2.1.3 Microwave. Microwave heating consists of heating the blade material with microwaves to prevent ice formation. The objective is to maintain the blade surface at a temperature slightly above 0°C, in order to save some energy that will be used for defrosting. It is recommended to cover the surface of the blade with a material that reflects microwaves (metallic material such as wire mesh) and apply paint to improve the final surface. Another method consists in heating the blades when they pass in front of the tower by fixing an emitter on the tower. It has been tested at the LM Glassfiber workshop on a LM19.1 blade with a 6-kW power, a frequency of 2.45 GHz and an emitted power less than 0.01 W/m^2 but is still to be implemented commercially.

4.2.2 Active de-icing systems

The active de-icing systems are used to eliminate ice accreted on the blade using a heating resistance, hot air, flexible pneumatic boots and electro impulsive/expulsive devices.

4.2.2.1 Heating resistance. The electrical heating uses an electrical resistance embedded inside the membrane or laminated on the surface. The idea is to create a water film between the ice and the surface. Once a film is created, centrifugal forces will throw the ice away. Electrically heated foils can be heating wires or carbon fibers. Heating elements cover the leading edge area of the blade. The ice detector and blade surface temperature are used to control the operation of the heating system. Additional temperature sensors are installed to protect the blade from permanent damage induced by over-heating. Heating foil can be applied to most turbines. A system of 15 kW per blade has been used for a 600 kW wind turbine, corresponding to 1%–4% of annual production, depending on climate conditions. The minimum time to keep the heating on, after the icing event has passed, is usually about 15–30 min. Heating demand is almost linearly dependent on the temperature difference between the air and the blade surface. More energy is needed to de-ice the tip's leading edge than that of the hub's (3.5–3.9 times more). More energy is also needed to de-ice the tip's trailing edge than that of the hub's (2.6–2.9 times more) and to de-ice the lower surface rather than the upper (1.3–1.5 times more).

This simple method has been used successfully in the aerospace industry for many years. It has also been used in the wind industry since 1990. Heating power seems to be adequate except in the case of super cooled rain. Thermal efficiency is close to 100% because of direct heating. Energy demand does not increase with blade size. As an inconvenience, there are many commercially available products but none are mass produced. The technology is still at the prototype level because of the limited market. If one heater fails, it will cause major imbalance on the whole system. In some extreme icing cases, blade heating power was found to be insufficient. Icing of the run-back water at the edges of the heating elements occurs quite often. When the running water from the heating element area reaches a cold blade surface, it re-freezes and forms a barrier at the edges of the heating element. The edge barriers may grow towards the leading edge as 'horns' without a contact with the heating element. This could explain why in some blade icing cases, the thermostat of the ice prevention system indicates a temperature higher than 0°C on the surface of the heating element during icing. Heating elements can attract lightning but lightning protection is efficient and no damage was detected in the ice prevention system studied by Marjaniemi and Peltola [9].

Advantages:

- Comparatively simple and widely spread.
- If considering the profitability of wind energy production the needed energy is small.
- Thermal efficiency is close to 100% due to direct heating.
- Energy demand does not increase with the turbine capacity increase.
- Comparatively low energy consumption.

Disadvantages:

- Despite the product availability there is no mass production.
- The technology prototype level because of the limited market.
- Even if only one heating element fail, the whole system may come into imbalance.
- Run-back water should be managed. If not it can reach the cold area of the blade and refreeze. Further, it can lead to the 'horn' appearance on the blade.

4.2.2.2 Electro impulsive/repulsive devices. This consists in very rapid electromagnetically induced vibration pulses in cycles that flex a metal abrasion shield and crack the ice. A spiral coil is placed near the surface of the blade. When a current is applied to the coil, a magnetic field is created between the coil and the thickness of the blade. The result is a rapid movement of the surface and the expulsion of the accumulated ice. The method has been recently certified for use on Raytheons Premier I business jet. It is used by Hydro-Quebec for transmission lines and Goodrich is currently developing this method for aeronautical applications.

Advantages:

- The system is efficient.
- Environmental friendly.
- Low-energy consumption.
- No interference with Hertz transmission.
- Easily automated.

Disadvantages:

- New technology that has not been tested on wind turbines.
- Ice expulsion is a potential problem.

4.2.2.3 Flexible pneumatic boots. Flexible pneumatic boots remove ice from the blade surface by breaking after surface inflation. After the buildup of about 6–13 mm of ice on the surface of the aerofoil, de-icers are inflated with compressed air. The inflation cycle lasts for a few seconds to achieve optimal ice shed and prevent additional ice formation on the inflated surface. After the ice has cracked, its bond to the surface is broken and it is removed through centrifugal and aerodynamic forces. Deicers are then allowed to deflate. Vacuum is then applied to ensure that there is no lifting of the surface on the low-pressure side of the aerofoil. De-icers for wind turbine applications have equivalent ice shedding and residual ice performance as conventional aircraft de-icers. Working at higher pressures for wind turbine applications, tests indicated satisfactory icing shedding on glaze ice at temperatures above –10°C and residual ice at temperatures between –10°C and –20°C.

Advantages:

- Low-energy consumption.
- Good results in laboratory tests.
- Low-energy consumption.

Disadvantages:

- Field-landed tests have to be made.
- Aerodynamic characteristics can be affected because the drag coefficient is increasing.
- Maintenance requirements during lifelong period can be expensive.
- High centrifugal loads at the outer radius of the pneumatic system will inflate itself or has to be divided into short sections.
- Detached ice fragments hazard should be taken into account.

4.2.2.4 Warm air and radiators. The main idea is to create a water film between the ice and the surface like a heating resistance, but instead of using an electrical resistance warm air is blown inside the blade. The blowers are located in the blade root; the internal blade volume is divided into two sections. These sections are used to guide a re-circulating hot air stream passing through the rotor blade. From the blower, the heated air flows directly along the blade's leading edge profile to the blade tip and then back to the blade flange. The returning air is then reheated and passed into the rotor blade. This closed circuit reduces the needed energy, because the initial air temperature is higher than the outside one. The process efficiency can be improved by using hot waste air from the machinery.

The ENERCON company has tried this method in its WEC. An evaluation of the efficiency of the rotor blade de-icing system was performed during the winter 2009/2010 at a wind farm site in northern Sweden (Dragaliden) and in Czech Republic (Krystofovy-Hamry). On each site, two ENERCON E-82 2.0 MW wind energy converters with hub heights of 108 m (Dragaliden) and 78 m (Krystofovy-Hamry) were used for the evaluation. Each WEC was equipped with the ENERCON rotor blade de-icing system, while only one turbine on each site operated with the system activated. The installed rotor blade de-icing systems had a power consumption of 23.8 kW per blade.

On both sites, the WEC with the rotor blade de-icing system in operation produced significantly more energy than the corresponding WEC. The mean deviation in the production amounts to about 50% and 46% of the production of the WEC without blade de-icing system in case of the test sites Dragaliden and Krystofovy-Hamry, respectively.

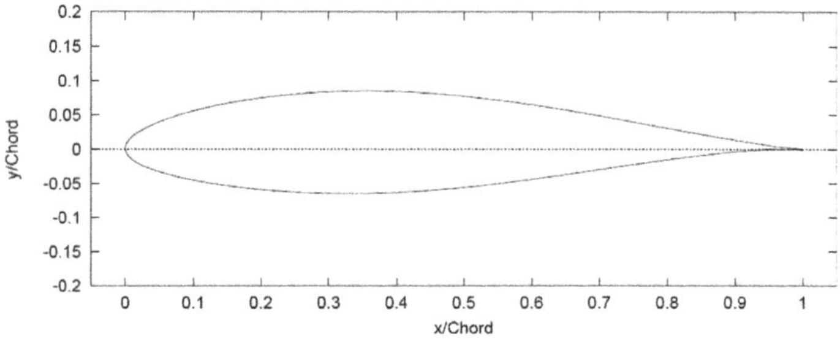

Figure 6: NACA 63415 profile [10].

5 Computational Analysis of Ice Accretion

After having described the theoretical aspects of ice accretion in wind turbines, this section will focus on the computational analysis of an icing event. All the simulations are performed by using the software called LEWINT, from American Kestrel Company. For this analysis, the input variables of an icing event are needed. Simulations are carried out for the relevant ranges of the variables.

It will be presented how a profile of an aerofoil for wind turbines reacts in an icing event. How ice is formed, the heat necessary to melt it and how much time is needed for the ice to be formed will be presented. This information is necessary to determine as a de-ice system has to be activated.

In this chapter, the effect of the rotational speed of the blades is not taken into account.

The output data extracted from the simulations are compared with some experimental data provided from other references. It has been found that the order of magnitude is similar to the experimental data. The difference between the simulations and the experimental data might be due to the effect of the blade's rotational speed and the consideration of the variable MVD.

5.1 Icing conditions

A simulation of an icing event needs all the input data relevant for such situations. In this chapter, the aerofoil profile considered is a NACA 63-415 because it is difficult to obtain industrial profiles as companies keep this kind of information secret. This profile is used in the industry of wind turbines. Figure 6 shows the profile of the NACA aerofoil.

The other variables being considered are relative humidity, air temperature, angle of attack, wind speed, median volume diameter (MVD) of the water droplets (MVD), LWC in the air, chord length and time of the icing event.

The AOA is the angle between the chord line of the aerofoil and the vector representing the relative motion between the body and the fluid flux, in this case the air, the unit is degree (°).

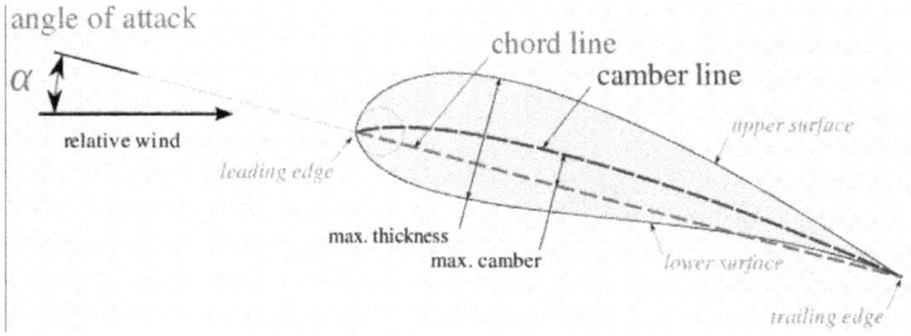

Figure 7: Characteristics of an aerofoil [12].

Table 1: Ranges for input data [2].

NACA airfoil profile	63415
Relative humidity (%)	100
Air temperature (°C)	–20 to 0
Wind velocity (m/s)	10–30
MVD (μm)	10–100
LWC (g/m^3)	0–2
Chord length (m)	1.435
AOA (deg.)	0–25
Time of icing (h)	0.5–5

Table 2: Values for input data.

NACA airfoil profile	63415
Relative humidity (%)	100
Air temperature (°C)	–2, v10, –20, –30
Wind velocity (m/s)	10, 30
MVD (μm)	15, 50, 90
LWC (g/m^3)	0.5, 2
Chord length (m)	1.435
AOA (deg.)	0, 5, 7
Time of icing (h)	4

The LWC is a measure of the mass of the water in a cloud for a specified amount of dry air. It is typically measured in grams per volume of air (g/m^3).

The MVD refers to the midpoint droplet size, where half of the volume of the spray is in droplets smaller and half of the volume is in droplets larger than the median. This variable is used instead of multiple drop size distributions. The difference in ice shape and icing limits is very slight for single body simulations.

The chord length is the length of the chord line and is the characteristic dimension of the aerofoil section. Of course, the chord length depends on the position of the aerofoil from the centre of the wind turbine.

The relative humidity (RH) normally is assumed to be 100%, unless these data are known. In this case, there are no data provided so it will be assumed to be 100%.

The analysis will take into account a total of nine input parameters. However, humidity and chord length will have just a single value and a specific NACA profile will be studied.

The other variables have different values depending on which icing event being evaluated.

- Chord length for an airfoil of 2 MW of power should be 1.435 m [11].
- The range of atmospheric temperatures will be from –30°C to 0°C.
- RH of 100%.
- The wind speed is from 10 to 30 m/s.
- The AOA is from 0°C to 25°C.
- The MVD has values from 10 to 100 μm.
- The LWC is from 0 to 2 g/m^3.
- The time of the simulated icing event should be no less than 30 min, because it is enough for the ice to form and to find out how it grows. As described the icing events can take hours, so an appropriate time simulation can be 4 h.

Figure 7 shows different aspects of an aerofoil profile and Table 1 shows the the range of the input data being analysed.

Taking into account the number of variables involved, it is necessary to restrict them to limited values representing all parts of the ranges. Due to the limitations of the software the AOA can not be higher than 7, because there are no data available in the program to simulate such conditions. Also MVD can not be higher than 500 μm for the same reason.

Finally, the values of the variables considered for the general analysis are shown in Table 2.

The time for general simulations is 4 h because this is sufficient to detect the evolution of ice accretion. Of course, in many studied cases the de-icing systems that melt the ice must act before this time because the final thickness is far from the acceptable maximum.

For the general analysis, the range of MVD is between 10 and 100 μm, although it can have higher values. However, for the dependence of ice thickness on MVD, the full range is taken into account.

5.2 Case studies

The simulations in this study provide the input data required to calculate the power loss occurring in each icing event. Also, by the ice thickness versus the time relation, the data necessary to decide when the de-icing systems have to start are provided.

The icing conditions calculated are based on weather prediction models. Then, the corresponding ice accretion rate and the amount of ice on wind turbine rotor blades are calculated using the icing simulation software LEWINT.

Then CFD simulations are used to get the degraded aerodynamic properties of blades and finally power performance simulations are used to generate the power curves. Measurement data are used to validate the power production simulations. The power production will be calculated based on wind data and representative power curves for the iced upwind turbine. For simplicity in the simulations, the icing is considered to be rime icing. Because of this, an additional degradation in power production has to be considered if the temperature on the site of interest is commonly in the range where glaze icing occurs, because it is known to have more deteriorating effects than rime icing. However, the focus is on the ice accretion rate and the amount of ice on wind turbine rotor blades.

A comparison on how different weather conditions generate different ice shapes is provided.

The aims of these simulations are:

- General analysis of the ice thickness and ice shape in different icing events and dependence on different variables.
- Comparison of the time required for different icing events to generate ice thickness.
- Analysis of the heat demand for some icing events.

5.3 Results

A general analysis of the ice thickness and ice shape for different icing events and dependence on the different variables is given in this section. The highest ice thicknesses of an icing event are compared. Also the dependence of ice thickness on the different variables is analysed.

1. Wind speed of 10 m/s, LWC of 0.5 g/m^3 and MVD of 15 μm:

Table 3: Ice thickness in millimetres.

AOA	Temperature (°C)			
	–2	–10	–20	–30
0	0	0.2	4.6	8.2
5	0	1.7	2.3	3.8
7	0.2	0.2	0.2	0.2

It can be seen from Table 3 that ice accretion is higher for lower temperatures. Also it seems that the AOA does not affect the ice thickness.

2. Wind speed of 10 m/s, LWC of 0.5 g/m^3 and MVD of 50 μm:

Table 4: Ice thickness in millimetres.

AOA	Temperature (°C)			
	–2	–10	–20	–30
0	13.8	39.4	40.3	41.2
5	16.8	38.7	39.6	40.4
7	20.0	38.0	39.1	39.6

3. Wind speed of 10 m/s, LWC of 0.5 g/m^3 and MVD of 90 μm:

Table 5: Ice thickness in millimetres.

AOA	Temperature (°C)			
	–2	–10	–20	–30
0	14.7	50.3	51.4	52.1
5	32.4	50.1	51.3	52.0
7	36.7	50.7	51.7	52.2

Tables 4 and 5 show much higher thickness than Table 3. This indicates that the ice thickness may have a great dependence on the MVD. However, 15 mm is a really small value, so maybe this value does not have a real significance.

4. Wind speed of 10 m/s, LWC of 2 g/m^3 and MVD of 15 μm:

Table 6: Ice thickness in millimetres.

AOA	Temperature (°C)			
	–2	–10	–20	–30
0	16.4	35.5	40.9	48.8
5	1.2	11.3	12.4	13.3
7	0.4	0.4	0.4	0.4

5. Wind speed of 10 m/s, LWC of 2 g/m^3 and MVD of 90 μm:

Table 7: Ice thickness in millimetres.

AOA	Temperature (°C)			
	–2	–10	–20	–30
0	22.5	98.0	148.2	195.1
5	33.8	146.8	165.0	193.1
7	34.2	165.8	161.8	186.1

Tables 6 and 7 show a great difference between various MVDs. The first one is much lower than the second one. The values of the ice thickness in Table 7 are high, and in real situations the wind turbine would be losing a lot of power output.

6. Wind speed of 30 m/s, LWC of 0.5 g/m^3 and MVD of 15 μm:

Table 8: Ice thickness in millimetres.

AOA	Temperature (°C)			
	–2	–10	–20	–30
0	31.4	65.3	68.1	69.2
5	26.5	48.1	52.5	58.0
7	15.4	30.1	38.0	39.2

7. Wind speed of 30 m/s, LWC of 0.5 g/m^3 and MVD of 90 μm:

Table 9: Ice thickness in millimetres.

AOA	Temperature (°C)			
	–2	–10	–20	–30
0	54.5	155.2	161.1	163.2
5	84.5	151.9	158.7	160.3
7	82.8	151.6	158.6	160.7

Tables 8 and 9 have a similar evolution as Tables 6 and 7. But for low temperatures, Tables 8 and 9 have larger ice thickness than those in Tables 6 and 7. For example, by comparing Tables 7 and 9 it can be seen that when the temperature is –2°C, the ice thickness is more than twice in Table 9 compared with that in Table 7,

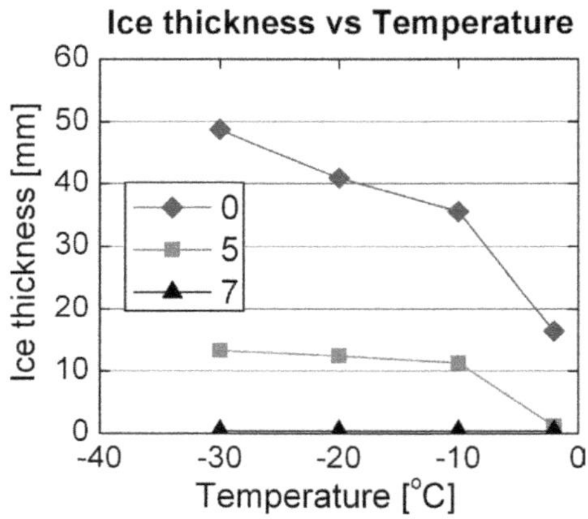

Figure 8: Wind speed 10 m/s, MVD 15 μm and LWC 2 g/m^3.

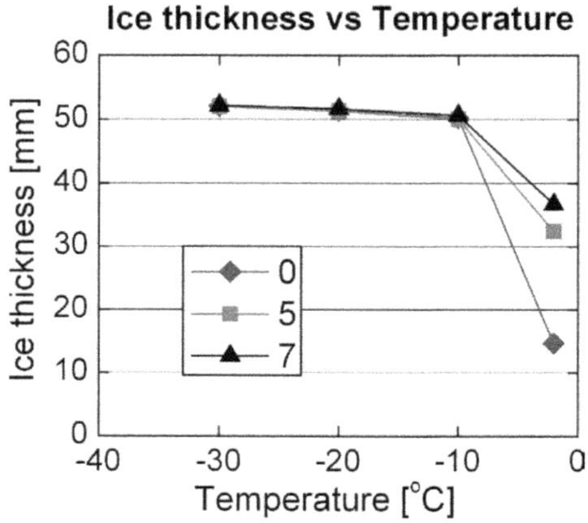

Figure 9: Wind speed 10 m/s, MVD 90 μm and LWC 0.5 g/m^3.

but when the temperature is –20°C the values are more or less the same. This may indicate that the wind speed has a stronger effect when the temperatures are close to zero than they are very low.

8. Wind speed of 30 m/s, LWC of 2 g/m^3 and MVD of 15 μm:

Table 10: Ice thickness in millimetres.

AOA	Temperature (°C)			
	–2	–10	–20	–30
0	36.4	106.1	219.7	206.0
5	66.1	105.3	136.3	152.5
7	8.7	115.1	144.8	142.6

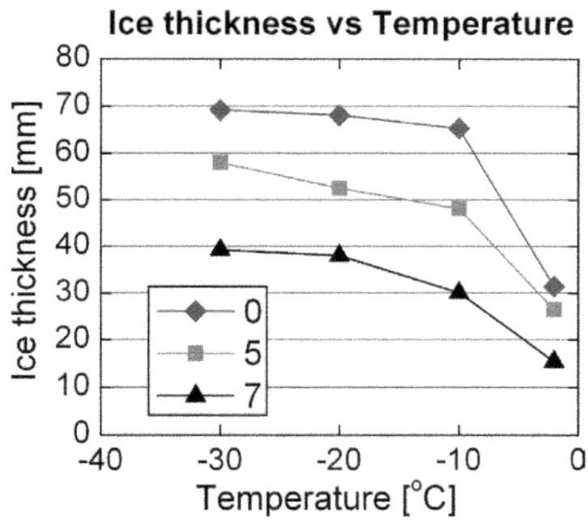

Figure 10: Wind speed 30 m/s, MVD 15 μm and LWC 0.5 g/m^3.

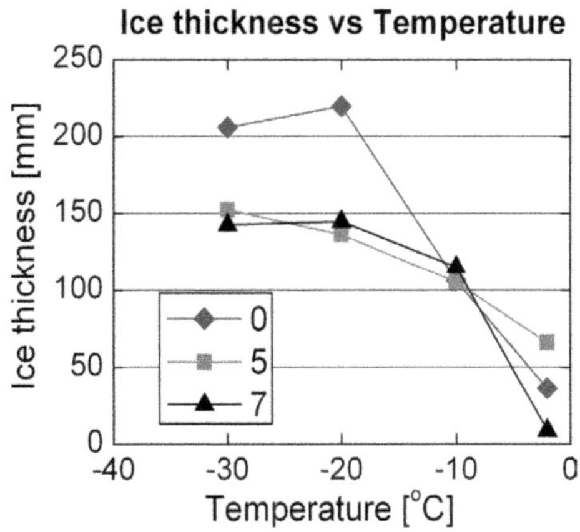

Figure 11: Wind speed 30 m/s, MVD 15 μm and LWC 2 g/m^3.

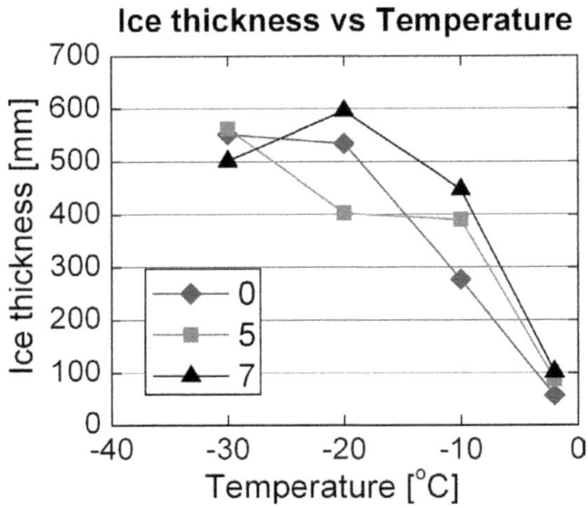

Figure 12: Wind speed 30 m/s, MVD 90 μm and LWC 2 g/m^3.

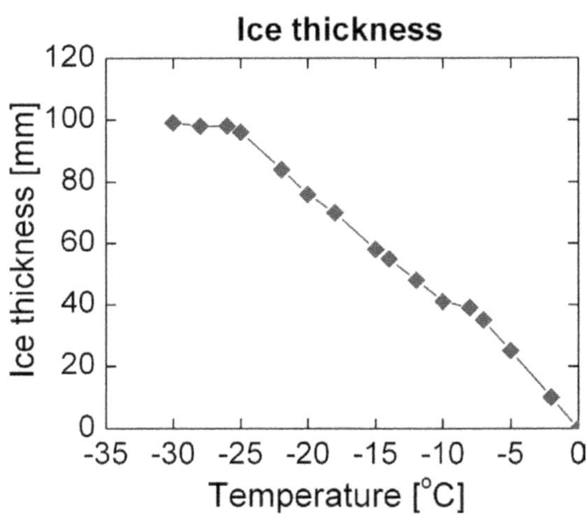

Figure 13: Dependence of ice thickness on air temperature.

9. Wind speed of 30 m/s, LWC of 2 g/m^3 and MVD of 90 μm:

Table 11: Ice thickness in millimetres.

AOA	Temperature (°C)			
	–2	–10	–20	–30
0	56.6	277.0	534.0	551.1
5	88.1	390.6	402.9	562.0
7	102.1	446.8	596.2	501.0

It has been found that among the variables that affect the ice accretion, the air temperature and wind velocity are the most important. Now some figures about the dependence of ice thickness on air temperature will be given. Also there are some aspects that it is not possible to show because of the limitations of the software. For example, the air density is a constant, but in reality it depends on air temperature. Also, the ice density is another limitation of the model, because it is assumed constant, but in reality it is not. These differences may change the ice thickness but not the evolution of it.

Figures 8–12 show the evolution of ice thickness with temperature by using the data from Tables 6, 5, 8, 10 and 11, respectively.

The data in Tables 3–11 and Figs. 8–12 provide a general view about the thickness of the ice that accretes on the aerofoil profile. It can be seen that with low temperatures the ice thickness becomes higher. Also for higher values of the wind speed, MVD or LWC the ice thickness seems to become bigger. Finally, the AOA does not affect the ice thickness to any significant amount.

Low temperatures have more energy potential due to higher air density. It is important that wind turbines can operate in such conditions, and then a de-icing system to melt the higher amount of ice is needed.

The ice thickness produces a power loss, also as the ice thickness causes the rotational speed to accelerate. This will damage the generator, as well as the electric equipment.

There are practical experiments, which show that with an ice thickness of 12.5 mm the power output is 15% of the regular one. Of course, it depends on the aerofoil, but it is an example of how power is affected by ice accretion [13].

After having presented a general view of the ice accretion, it is clear that the variable affecting most significantly is the air temperature, as is evident in Tables 3–11 and in Figs. 8–12. However, in the next subsections, the dependence of the ice thickness on all variables is analysed.

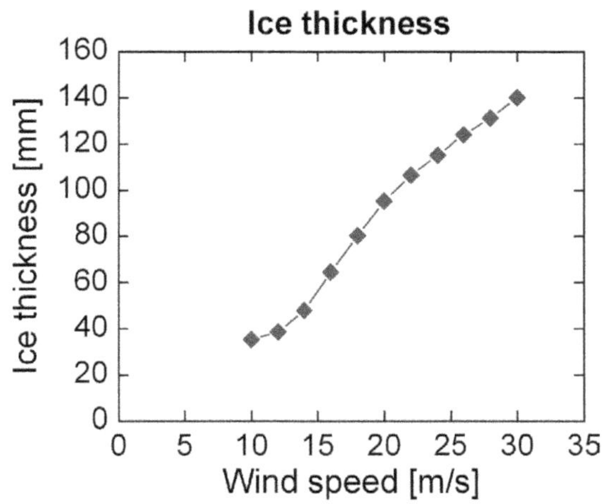

Figure 14: Dependence of ice thickness on wind speed.

5.3.1 Dependence of ice thickness on air temperature

By comparing Tables 3–11, it is clear that the effect of air temperature is higher when the temperature is low. However, it can be seen that in most of the cases there are two different zones. The first zone is in the range between 0° and –10°. The characteristic of this zone is that the ice thickness grows rapidly. On the other hand, in the range between –10° and –30°, the ice thickness seems to be more stable and the growth is less. But this difference might be due to the combination of the different variables.

To make a more accurate study of the dependence of the ice thickness on temperature, all the other variables are fixed and the 16 different values of the air temperature are evaluated.

The values of the different variables are: wind speed 10 m/s, LWC 2 g/m^3, MVD 90 μm and AOA 0° and time simulation 7200 s.

As can be seen in Fig. 13, despite the first impressions, the dependence of the ice thickness on the air temperature is quite regular in the whole ranges of the variables. However, the difference between the ice thickness from 0° to –10° is higher than the one between –10° and –20°, and the latter one is higher than the one between –20° and –30°. As shown in Fig. 13, the dependence of ice thickness on temperature is almost linear.

By comparing icing conditions, the ice thickness is greater at low temperatures. The significant difference in ice thickness is because of the different value of the collection coefficient. Low temperatures have higher collection coefficients.

5.3.2 Dependence of ice thickness on wind speed

In general, it seems that with higher velocities, the ice thickness is higher.

The values of the different variables are: MVD 100 μm, LWC 1 g/m^3, temperature –10° and AOA 0° and simulation time is 7200 s.

The operational wind velocity of wind turbines ranges from 10 to 30 m/s, and, as it can be seen in Fig. 14 that the ice thickness is higher when the wind velocity is higher. The dependence seems to be linear as well.

5.3.3 Ice shape

The ice accretion events have been compared only by taking into account the greater thickness for each case. However, the shape of the aerofoil after icing events is also important because the aerodynamics are affected.

Aerodynamic characteristics of the wind turbine blades including lift and drag coefficients are affected by the ice accretion, and accordingly the power produced is different for each icing event.

The ice shape is also important. The de-icing systems focus on a certain part of the aerofoil. If it is known where ice forms the amount of heat produced to melt the ice will be less.

6 Conclusions

The topic of this chapter is really new and it is important nowadays because of the benefits of cold climates for wind turbines. This means that there is not much data available about icing events. Thus, it has been difficult to obtain the information needed. This chapter resumes part of the field of ice accretion in wind turbines. It shows different icing events, the weather variables of importance and the effects of ice accretion.

Also systems of anti-icing and de-icing are described. As it is a relatively new field not much work has been done on such systems. Accordingly, a lot of work is needed to optimise these processes. Most of these copy the systems installed in aircraft, because this field has more or less the similar problems, but in different situations. Warm air, a specific de-icing system, is described in details because there was information provided by the ENERCON company. It shows two different places with icing events and the difference of having a de-ice system and not having it is really important. The difference of having a de-ice system is a gain of 50.4% in power production.

There is a general physical model for ice accretion but it has some limitations that have to be corrected. One of them is that there are some coefficients in the formula that are based on cylinder shapes. Some programs transform these shapes to more complex structures like blades, but it would be better to reformulate the model.

The simulations show differences in the dependence on ice accretion of the different variables. The ambient temperature is the most important factor to take into account for icing events. The wind speed is the second most important variable. Also, if the rotational speed is added the relative wind speed in different parts of the blade would make this variable even more important. LWC is also important, but depending on the range of its value it does not really affect the ice accretion. The same is true for MVD, if the values of this variable are low the ice thicknesses of each case are really different, but after 100 μm the effect of it seems to be not important. Finally, the effect of AOA on ice thickness seems to be insignificant. However, AOA affects the ice shape, so it is a variable that must be taken into account depending on which kind of de-icing systems is installed in the blade turbine, for example heating resistance, that focus on a specified part of the blade surface.

Acknowledgement

Financial support from the Swedish National Research Council and the Swedish Energy Agency is gratefully acknowledged. Most of the work was carried out as a Master thesis project by Mr. J. Pallarol at Lund University.

Nomenclature

A	Area the wind is passing perpendicular to the wind (m^2)
b	Radiation linearisation constant (K^3)

C_{pa}	Specific heat of air (J kg^{-1} K^{-1})
C_{pw}	Specific heat of water (J kg^{-1} K^{-1})
Cp	Constant efficiency for wind turbines
D	Cylinder diameter (m)
d	Droplet diameter (m)
d_m	Droplet median diameter (m)
e_s	Saturation water vapour pressure (Pa)
e_a	Ambient vapour pressure in the airstream (Pa)
h	Heat transfer coefficient (W m^{-2} K^{-1})
L_e	Latent heat of vaporisation (kJ kg^{-1})
L_f	Latent heat of freezing (kJ kg^{-1})
P	Power produced from wind turbines (W)
P_w	Power available in the wind (W)
p	Air pressure (Pa)
R	Radius of the rotor (m)
Re_d	Reynolds number
t_s	Surface temperature (K)
V_{tip}	Tip speed (m s^{-1})
w_r	Rotational speed of the rotor (s^{-1})
υ	Free stream velocity (m s^{-1})
υ_p	Particle velocity relative to the object (m s^{-1})
υ_o	Droplet impact speed (m s^{-1})
υ_w	Wind velocity (m s^{-1})
ω	Mass concentration (kg m^{-3})
λ_{sr}	Tip speed ratio
μ	Viscosity of air (Pa s)
ρ_a	Air density (kg m^{-3})
ρ_{ice}	Ice density (kg m^{-3})
ρ_w	Water density (kg m^{-3})
σ	Stefan–Boltzmann constant (W m^2 K^4)

Abbreviations

ADIS	Anti-icing and de-icing systems
AOA	Angle of attack
CC	Cold climate
CFD	Computational fluid dynamics
IC	Icing climate
LTC	Low-temperature climate
LWC	Liquid water content
MVD	Median volume diameter
RH	Relative humidity
WEC	Wind energy converter

References

[1] Sanderud, S., *Estimating Production Loss due to Icing on Wind Turbines*. Master Thesis, Department of Physics and Technology, University of Tromsø, Norway, 2011.

[2] Laakso, T., Baring-Gould, I., Durstewitz, M., Horbaty, R., Lacroix, A., Peltola, E., Ronsten, G., Tallhaug, L. & Wallenius, T., *State-of-the-Art of Wind Energy in Cold Climates*. VTT-WORK-152, VTT Technical Research Centre of Finland, Finland, 2010.

[3] Rindeskär, E., *Modelling of Icing for Wind Farms in Cold Climate*. Master Thesis, Department of Earth Sciences, Uppsala Universitet, Sweden, 2010.

[4] Baring-Gould, I., Cattin, R., Durstewitz, M., Hulkkonen, M., Krenn, A., Laakso, T., Lacroix, A., Peltola, E., Ronsten, G., Tallhaug, L. & Wallenius, T., *Wind Energy Projects in Cold Climates*. International Energy Agency, 2011.

[5] Ilinca, A., *Analysis and Mitigation of Icing Effects on Wind Turbines*, ed. I. Al-Bahadly, Wind Turbines. InTech, 2011.

[6] Makkonen, L., Models for the growth of rime, glaze, icicles and wet snow on structures. *Philosophical Transactions of Royal Society of London*, **358**, pp. 2913–1939, 2000.

[7] Marmytova, S., *Wind Turbine Operation in Cold Climate*. Master Thesis, Degree Programmme in Energy Technology, Faculty of Technology, Lappeenranta University of Technology, Finland, 2011.

[8] Parent, O. & Ilinca, A., Anti-icing and de-icing techniques for wind turbines: critical review. *Cold Regions Science and Technologies*, **65**, pp. 88–96, 2011.

[9] Marjaniemi, M., & Peltola, E., A blade heating element design method. *In Proceedings of Wind Energy Production in Cold Climates*, BOREAS III, Finnish Meteorological Institute, 1996.

[10] Bertagnolio, F., Sörensen, N., Johansen, J. & Fuglsang, P., *Wind Turbine Airfoil Catalogue*, Risö-R-1280, Risö National Laboratory, Denmark, 2001.

[11] Homola, M., Wallenius, T., Makkonen, L., Nicklasson, J. & Sundsbø, A., The dependence of icing severity on chord length. In EWEC, Marseille, France, March 16–19, 2009.

[12] Wikipedia. *Airfoil*. http://en.wikipedia.org/wiki/Airfoil.

[13] Wang, X., *Convective Heat Transfer and Experimental Icing Aerodynamics of Wind Turbine Blades*. Master Thesis, Department of Mechanical and Manufacturing Engineering, University of Manitoba, Canada, 2008.

CHAPTER 7

Structural Consideration for Wind Turbine Blades

R.S. Amano
University of Wisconsin-Milwaukee, Milwaukee, WI, USA.

Abstract

This chapter states the methods for a blade breakage self-repair system with an innovative self-healing approach. To prevent blade breakage due to excessive wind speed, the methodology of self-healing technology can possibly be implemented with a self-healing wind turbine blade – a new and innovative concept for wind turbine blade design. The composite blades are manufactured using a laborious, hand lay-up technique with glass fibre reinforcements. The use of fibre-reinforced composite materials has grown rapidly and such composites are often used in aerospace and other applications; however, concerns remain about the structural integrity of composite materials following impact loading, as they are susceptible to cracks or delaminations that form deep within the structure. With the success of the self-healing technology for wind turbine blades, any cracked parts in wind turbine blades can be healed during operation without any system shutting down. Therefore, technology sheds light on the existing wind turbine failure prevention methods and provides new concepts and approaches for new materials of turbine blades suitable for the 21st century wind energy era.

Keywords: Turbine blade failure, repair system, self-healing technology of turbine blades.

1 Introduction

The worldwide demand for energy is expected to double by the year 2030 and triple by 2050, when fossil fuels will account for no more than two-thirds of all energy consumed, compared with 79% of the energy consumed today [1]. Traditional fossil sources such as oil, gas and coal are not renewable and cause pollution by releasing huge quantities of carbon dioxide and other pollutants into the atmosphere, thereby damaging the environment in many ways, from acid rain to climate change. To help combat these problems, many states in the United States are seeking ways to use renewable energy sources, such as wind, solar and biomass. Along with its environmental and cost benefits, renewable energy is a rapidly growing industry with vast potential for economic growth and job creation. In fact, the U.S. Secretary of Agriculture has identified wind, solar and biomass as key factors for advancing the U.S. economy [2].

Wind energy has recently become the world's fastest growing source of renewable energy. The U.S. Department of Energy (DOE) expects that wind energy will contribute to 20% of the U.S. electricity supply by 2030. As a result, there has been a revived interest in wind turbines because they are emissions-free and wind is renewable and cost-free; however, the amount of electricity generated and obtained by wind energy conversion systems is still unsteady, relatively expensive and difficult to integrate into traditional electricity systems because of the variation in wind source and unresolved energy storage issues. On a large scale, spatial variability describes the fact that there are many different climatic regions in the world, some much windier than others. These regions are largely dictated by the latitude, which affects the amount of insolation. Within any one climatic region, there is a great deal of variation on a smaller scale, largely dictated by physical geography – the proportion of land and sea, the size of land masses and the presence of mountains or plains, for example. The resource map of wind energy in the United States [3] indicates the vast majority of available wind is very unsteady; strong wind zones are concentrated in certain regions and not uniformly distributed throughout the nation, making wind power collection more difficult. Conversely, the easy-to-collect wind energy is primarily confined to remote locations, making electricity distribution difficult. To achieve economic feasibility in enhancing the power in low- to high-wind areas, significant advances in turbine design are necessary for increased power collection. This chapter presents two major topics: (1) developing self-healing concept implementing into a wind turbine blade design and (2) failure prevention/self-repair of wind turbine blades during strong winds.

2 Wind Turbine Blade Failures

Turbine blades for the three-blade, horizontal-axis wind turbine (HAWT) model for 500 kW to 2 MW sizes are traditionally designed with a straight blade, which has significant interactions with the terrain and the other wind turbine units. Therefore, more in-depth research is required to understand the operation and the mechanisms of the HAWT units. The PI's group at the University of Wisconsin-Milwaukee (UWM) has investigated power performance on different wind turbine blade designs for many years and recently discovered a new structured design of blades that shows remarkably improved life performance [4].

The implementation of a self-healing mechanism is one of the effective methods to prevent turbine blade damage, because breakage significantly affects a steady power generation operation, as well as local residents near wind farms who might suffer from damaged blades. For example, on 14 April 2005, the operators of Crystal Rig Wind Farm in the Lammermuirs investigated why a 40-m blade from one of its state-of-the-art turbines suddenly shattered (Fig. 1(a)). Another incident on 4 September 2008, in Northern Ireland also shows turbine blade breakage occurring at nearly a similar location in the blade span (Fig. 1(b)). In both cases, the wind speed was a little over 20 m/s. The PI's group performed both computational fluid dynamics (CFD) and FEA analysis and discovered aerodynamic high stress at the area where the blade snapped off (Fig. 1(a) and (b)). As discussed in the PI's preliminary analysis (Fig. 1(c)), the highest pressure and maximum stress for the case with wind speed of 20 m/s occurred at the mid-section of the blade span, corresponding with the breakage part of two breakage incidence cases, as depicted in Fig. 1(a) and (b).

The fatigue damage analysed by Martin *et al.* is shown in Fig. 2 [5]. In their analysis, the cell of the trailing edge is closed in the zone of transition by means of an element triangular in shape, constituted by a laminate of low resistance, an element that we refer to as a cover. As can be observed in Fig. 2(a), the crack extends throughout both the union cover-root and the transition root-aerodynamic zone. A detail of the damaged area (Fig. 2(b)) at slice number 7 is shown in Fig. 2(c), whereas those areas corresponding with slice numbers 5 and 6 are shown in Fig. 2(d), giving us a good concept of crack propagation. Fibre-reinforced material installation in this area could strengthen the blade to prevent cracks.

There are new uses for polymers and structural composites in a wide variety of areas, including aerospace, automotive and space applications; however, such materials are prone to failure from a number of factors, including rupture and

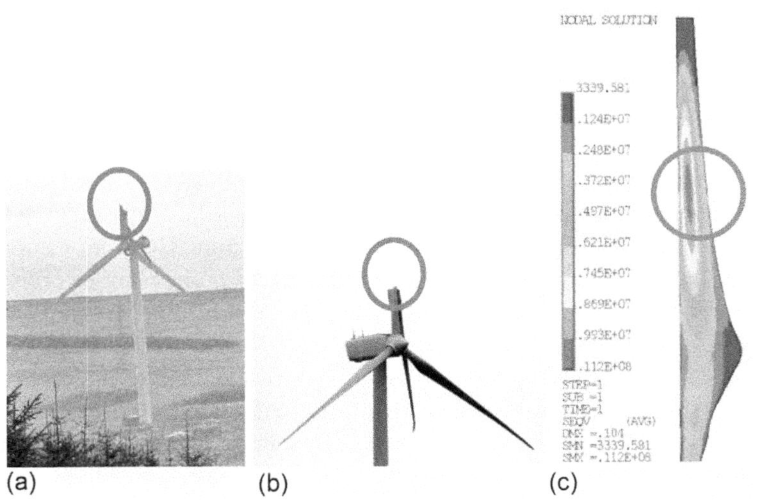

Figure 1: Turbine blades after failure and finite element analysis. (a) April 2005, Lammermuirs Hill, (b) September 2008, Northern Ireland and (c) FEA analysis by Amano's group [6].

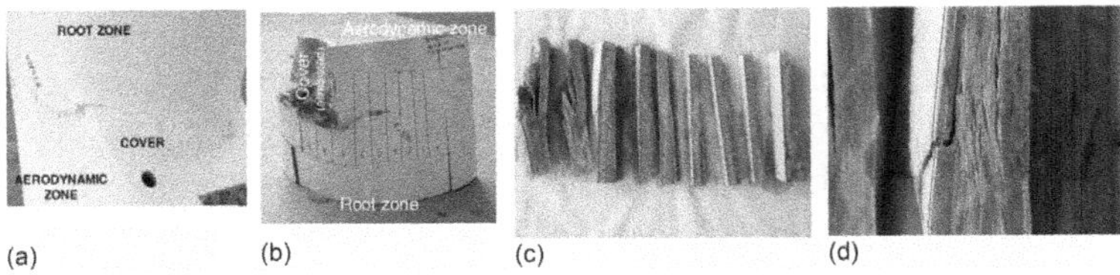

Figure 2: Wind turbine blade damage [9]. (a) Damage on a surface, (b) damage near the root, (c) progression of internal damage and (d) close-up of damage.

fatigue. These create micro-cracks in the materials that could propagate, thereby affecting either matrix-based or fibre-dominated properties of the composite. With polymers and composites being increasingly used in structural applications in aircraft, cars, ships, defense and construction industries, several techniques have been developed and adopted by industries for repairing visible or detectable damages on the polymeric structures. However, these conventional repair methods are not effective for healing invisible micro-cracks within the structure during its service life. In response, the concept of self-healing polymeric materials was proposed in the 1980s [7] as a means to heal invisible micro-cracks to extend the working life and safety of the polymeric components, which is a novel concept that can be used as an alternative to damage-tolerant design and temporary repairs to damaged structures.

3 Self-Healing Technology

3.1 Concept of self-healing

Nature's ability to heal has inspired new ideas and new mechanisms in the engineering community. Self-healing is defined as the healing of a damaged member with minimal human intervention – the autonomous healing of any member by itself. Self-healing is one of the many techniques mimicked from nature, and providing materials with the ability to self-heal cracks at the point of origin will improve the life and usability of these materials. Self-healing materials are no longer an illusion, and we are not far away from the days when manufactured materials can restore their structural integrity in case of failure [8].There is, in fact, increasing interest in developing synthetic, self-healing structural materials for applications to engineering components. To achieve this goal, understanding of self-healing in biological systems is important, as almost all biological systems exhibit self-healing. For example, human tissues and bones are able to self-repair when damaged; when a bone fractures, the healing process involves delivering material to the damaged site by blood vessels and bone marrow. The materials are deposited and assembled to bridge the gap and heal the bone. Human bone repair consists of material production, transport and assembly at the site of damage. A similar system has been developed in structural materials by placing some form of healing agent within the structural member. This healing agent is encapsulated in a hollow reinforcement such as a hollow sphere or a hollow fibre.

Chemists and engineers have proposed different healing concepts that offer the ability to restore the mechanical performance of the material. One area of interest is the fusion of the failed surfaces. Polymeric materials possessing selective cross-links between polymer chains that can be broken under load and then reformed by heat have shown to offer healing efficiencies of 57% of the original fracture load [10]. Another example presented by Zako and Takano is where a polymeric material hosts a second solid-state polymer phase that migrates to the damage site under the action of heat [11]. Hayes *et al.* [12] developed a two-phase, solid-state repairable polymer by mixing a thermoplastic healing agent into a thermosetting epoxy matrix to produce a homogeneous matrix that contrasts with the discrete particles of uncured epoxy reported by Zako and Takano. These systems offer the capacity for self-healing but require damage-sensing, some form of higher decision-making via a feedback loop, and heating requirements that are likely to be impractical in a real application, such as for wind turbine blades.

Wind turbine research has focused on developing stronger materials that will survive in strong wind conditions. The materials of structural wind turbine blades consist mostly of polymers such as carbon fibre/epoxy, carbon fibre/polyester or glass-reinforced thermoset. The non-metallic composites take in the impact loads by forming micro-cracks within the body, which propagate and result in failure. Many researchers have developed self-healing in ceramics and polymers, and because of this, the self-healing of polymer technology can be applied to a wind turbine. A research group at the University of Illinois Urbana-Champaign [13] developed self-healing polymers using microballoons, which were filled with a healing agent that leaked into the body when damaged and reacted with the embedded catalyst, causing polymerisation and crack healing. Figure 3(a) shows that a micro-capsulated healing agent is embedded in a structural composite matrix containing a catalyst capable of polymerising the liquid healing agent. When cracks occur, the crack ruptures the microcapsules, releases the healing agent into the crack plane and the healing agent contacts the catalyst, triggering polymerisation that bonds the crack. Since no heating is required, this technology is suitable for a wind turbine blade design.

3.2 Microcapsule self-healing

White *et al.* mimicked the bone marrow biological self-healing system to a certain extent by using microballoons filled with a healing agent [13] in a thermo set epoxy composite. A catalyst, diethylenetriamine (DETA), was embedded within the composite and a healing agent was encapsulated in microspheres. A propagating crack within the material breached the wall of the microballoon and caused the healing agent to flow into the crack (Fig. 3). The catalyst, which was already embedded in the material, reacted with the healing agent and was polymerised. The polymerisation filled the crack effectively, ultimately shutting the crack. The

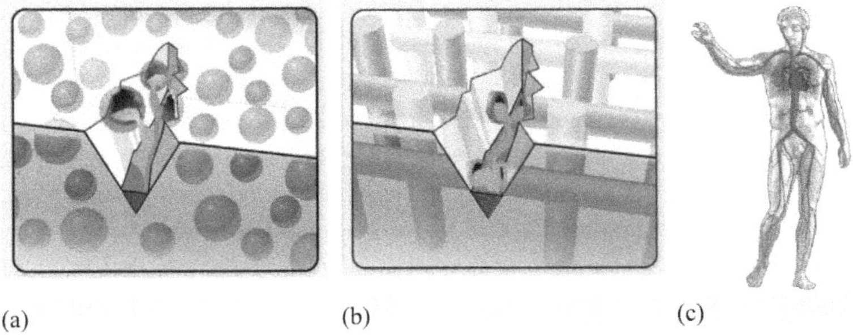

Figure 3: Self-healing concepts. (a) Microcapsule self-healing, (b) hollow fibre self-healing [15] and (c) networking [36].

microballoons used ranged in size from 50 to 200 μm. This method showed a 75% recovery in tensile strength after healing. Compared with the plain epoxy, the epoxy with micro-spheres and the catalyst showed a 20% increase in strength.

A key development in practical self-healing materials is the development of microencapsulated monomer healing agents. Microencapsulation, the process in which tiny droplets are enveloped by a polymer shell, is an established technology [16]. The small size of the spherical microcapsules enables the healing agent to be distributed throughout a matrix material to potentially heal microcracks before they coalesce into larger scale damage. Early work by Jung *et al.* [17] demonstrated the process by which polyoxymethylene urea microcapsules in a polyester matrix broke and released their content when encountered by a crack (see Fig. 3(a)); however, these systems were also limited by their chosen healing agent (in this case a mixture of styrene monomers and high molecular weight polystyrene) and showed no evidence of self-repair, but looked naturally maintained in an original shape. The microcapsule-based self-healing approach has the major disadvantage of uncertainty in achieving complete healing, because it has a limited amount of healing agent and it is not known when the healing agent will be consumed entirely.

3.3 Hollow-fibre self-healing

Another group at the University of Bristol, UK, created a self-healing matrix using hollow glass fibres that stored the healing agent before to crack breaching. A similar action can be done in hollow fibres, as shown in Fig. 3(b). Trask *et al.* [18] incorporated self-healing capability in fibre-reinforced polymers (FRP) by placing hollow glass fibres within glass fibre/epoxy and carbon fibre/epoxy laminates. A schematic of this method is shown in Fig. 3(b). The hollow glass fibres, ranging from 30 to 100 μm diameter and hollowness about 50%, were used to encapsulate the healing agent, Cycom 823, until healing was required.

Hollow glass fibres are preferred to microspheres, as they can integrate easily with the composite and act as reinforcement. The self-healing fibres are placed along with the regular plies at damage-critical interfaces; the glass and carbon FRPs had a similar arrangement. The tensile strength recovery was 87% in glass fibre-reinforced polymer (GFRP) and 80%–90% in carbon fibre-reinforced polymer (CFRP). These results are from a four-point flexural testing. The hollow glass fibres showed an advantage in the manufacturing process, as they can easily infiltrate with the healing agent by vacuum or pressure infiltration, while the microspheres must be infiltrated in situ during manufacture. To replenish healing agent, the hollow-fibre tubing must be arranged in a network, just like a human body (Fig. 3(c)).

To achieve self-healing, the propagating crack must rupture the microcapsule or hollow reinforcements. The relative toughness, interface strength between the reinforcement and the matrix, and wall thickness are several design parameters that need consideration for self-healing. A tough microcapsule or fibre will cause the crack to deviate from itself and the crack will go around it rather than rupture. Similar results are encountered in the case of a low interface bonding or a thick wall. A very thin wall also is undesirable, as the microcapsule or hollow fibre must hold the strength during the matrix forming; therefore, the materials used in the studies should be very case-specific. Any catalyst-activated resin system can be used for self-healing. Several important criteria in the material selection [19] include temperatures of the component, the curing temperature of the base matrix and the bonding between and with the matrices.

4 Scientific Background: Application Self-Healing for Wind Turbine Blades

As with any other component, wind turbine blades also suffer from stresses and resulting failures. Self-healing composites, though developed in the lab, have yet to be seen in the commercial market. To apply the self-healing method to a wind turbine blade, the design is based individually on the component. Wind turbine blades are a comparatively recent development, and incorporating self-healing in a wind turbine blade is very difficult to design in general, as no one has ever done this invention before. With improving blade designs and more complicated manufacturing processes, it is increasingly important that the blade does not fail during use. Self-healing can provide an effective method to prevent failures during operation and maintain the generation of electricity [20].

The future of the self-healing concept for composite materials with application to wind turbine blades relies on the development of a continuous healing network embedded within a composite laminate that delivers a healing agent from a reservoir to regions of damage to repair all types of composite failure modes. New and novel approaches must be found to ensure that the healing agent can be replenished and renewed during the life of the structure. The healing agent must restore the matrix material properties and restore the structural efficiency of fractured fibres. The initial steps towards a networking for any system have been taken, although a truly self-healing composite microvascular network still has not been attained like the human body (Fig. 3(c)). Toohey *et al.* [21] used an interconnected microvascular microchannel network in the coating to flow the healing agent throughout an epoxy polymer block. Conversely, Williams *et al.* [22] proposed and validated a simple vascular network within a composite sandwich structure consisting of channels approximately 1.5 mm in diameter, within a polymethacrylimide (Rohacell) core capped with glass fibre-reinforced epoxy skins. The generation of the network within the foam core offers a promising delivery system for healing conventional sandwich structures; however, the network is only one key area for advancement – without the corresponding development of a synthetic healing resin, the benefits of the delivery system will be lost [23].

Wind turbine blades are typically made of fibre-reinforced plastic, and the self-healing technique in such materials is a candidate for engineering applications; thus, the materials and reinforcements must be designed to suit the specific requirements of the wind turbine blades. Some experimental and computational models for self-healing material concepts have been attempted by several researchers [17,18,19,24]. The distributions of the stress and deformation within the blades are the major factors that must be determined to design self-healing reinforcements and to place them at the weak points, as addressed in the following sections.

5 Applications to Wind Turbine Blades

Several researchers have found success with a dicyclopentadiene (DCPD)–Grubbs catalyst combination to provide significant self-healing capabilities in epoxy-based composites (Fig. 4). The same combination of materials is used in this study, with the DCPD confined in the hollow glass tubes. The Grubbs catalyst was introduced into the matrix in two different forms: distributed as solid particles in the matrix and dissolved in dichloromethane and coated on the outer surface of the tubes.

Glass fibre-reinforced epoxy-based and polyester-based plastic specimens are one method to create through a vacuum resin infusion setup. Hollow glass tubes are inserted perpendicular to the glass fibre orientation without any DCPD. Once the crack is generated, the hollow glass fibre is filled with DCPD and enough curing time is given for the ring-opening metathesis polymerisation (ROMP) reaction to occur. To determine if the growing crack front breaks the glass or pass around it, an analytical study must be performed before the experiments to determine the geometric stress at the tip of the micro-cracks usually formed in GFRP.

- The main parameters of the determination are:
- Diameter and strength of glass tube.
- Self-healing efficiency in epoxy and polyester matrix.
- Tube DCPD pressure.

The efficiency of the self-healing assessed here by determining the amount of interlaminar fracture toughness recovered by self-healing is the key component for a successful design implementation.

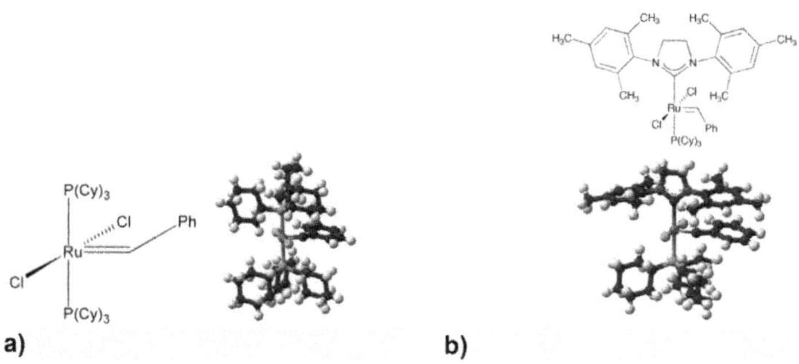

Figure 4: Grubb's catalyst (from Novel Lecture) [25]. (a) Grubb's first-generation catalyst and (b) Grubb's second-generation catalyst.

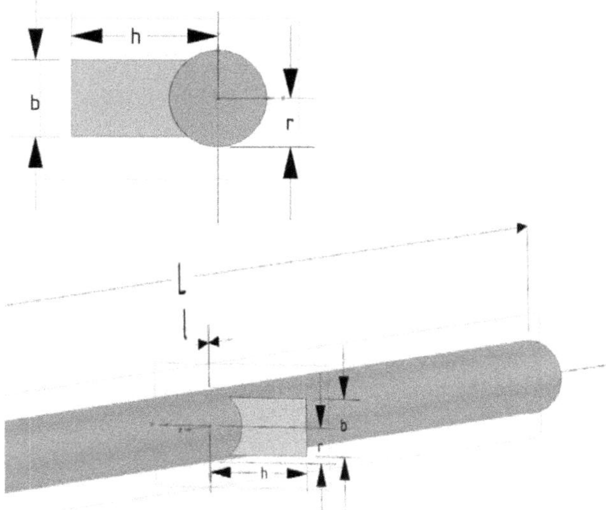

Figure 5: Crack and tube – relevant dimensions.

A CFD model is usually be created for determining the conditions under which the self-healing liquid flows from the tube into the crack (see Fig. 5). Then, the initial results are verified from experiments. The factors that have been identified by designers are:

- Size of the crack (length and half angle).
- Liquid properties – surface tension.
- Tube pressure.
- Relative volume (crack volume/tube volume).
- Open or closed crack.

Some of the results obtained from the CFD analysis include the flow time required for DCPD to fill the crack, extent of crack filling and the effect of the parameters mentioned earlier.

6 Blade Formation Process

6.1 Flow measurements

Experiments were carried out in a wind tunnel for scaled wind turbines to investigate the failure mechanism for rotating blades installed with a self-healing assembly. In the author's lab, a wind tunnel as shown in Fig. 6(a) was used. Measurements of velocity, pressure, stress and loading were performed in this test set (Fig. 6(b)), which gives uniform wind flow for steady data measurement for the velocity and pressure fields of the flow over the rotating wind turbine blades using both hot-wire anemometry (HWA) velocity measurements (controller through traverse system as shown in Fig. 6(d)) and a particle image velocimeter (PIV). The surface pressure of the blades can be measured through a slip-ring data acquisition transmission technology, which the PI's group has been using for more than a decade to measure flow on rotating fan-blades, lawn-mower blades and compressor blades [26].

The design step of the blade is summarised in Fig. 6(c). First, the blade is designed through an aero-elastic program and generated using a CAD program, which is followed by the CAD development procedure [27]. After these procedures, the stack-up of the blade shape is performed using a novel approach that combines the robust arithmetic for multivariate Bernstein-form polynomials and the Bezier surface segmentation algorithm [28]. The generated blade is then analysed using the CFD approach and the flow behaviours are investigated for a lower turbulence wake and higher lift coefficients states [29]. The resultant blade is then fabricated using the rapid-prototyping machine, and the test is conducted in the wind tunnel for flow and stress measurements. After analysis through the optimisation code, the data are the final candidate for manufacturing a new blade.

6.2 Blade formation

For a large blade formation that cannot be fabricated by a prototyping machine, a unique process constructs the blades, which was created because of the profile change of the blade. The profile changes made it nearly impossible to produce three identical blades by hand. Figure 7(a) shows the final construction of the blade. The blades were constructed out of fibreglass and expanding urethane structural foam. Figure 7(b) illustrates the materials and their locations within the blade structure. The core of the blade is a rapid expansion, two-pound density polyurethane foam; the next layer is made of three layers of fibre glass; and the last layer is a smoothing and finishing layer that creates a protective layer between the atmosphere and the fibre glass. For a small blade that is <25 cm, the rapid-prototyping machine was used (Fig. 7(c)).

First, the blade profile was created using Pro/E. Using that profile, a full-scale blade is made with a rapid prototype machine. Figure 7(e) and (f) shows the production of the blade in the rapid prototype machine. Next, the mould was created, which was used to create the other blades. The rapid-prototyped blade was then used as a plug for the blade mould; a parting line also is introduced to easily remove the finished blades. The rapid-prototyped blade is covered in five layers of

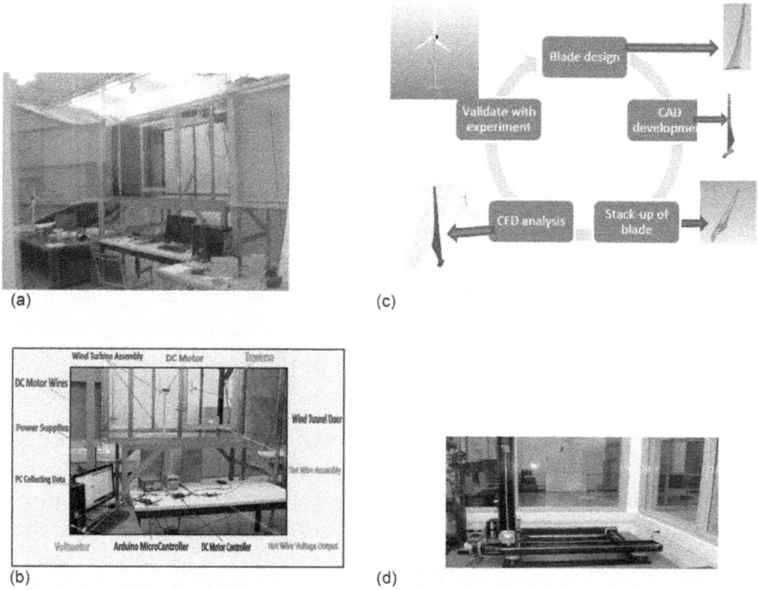

Figure 6: Experimental set up. (a) Wind tunnel for rotating blade measurement (b) measurement of flow velocities, pressure behind an airfoil and stress through the blade surface, (c) design process and (d) three-axis traverse and Velmex motor controller (UWM Wind Tunnel Lab).

fibreglass to add extra rigidity for extended use. The finished blade mould is shown in Fig. 7(e). Additionally, a specific fibre glass mould wax is applied to remove the blades easily from the mould. The fibre glass is pre-cut to the necessary shape and laid into the mould, and then wetted out with a polyester vinyl resin. The separate mould pieces were then clamped together and filled with the expanding urethane foam. The expanding foam internally exerts pressure upon the wetted-out fibreglass cloth and holds the wetted fibre glass in place, squeezing out any excess resin. Structurally, the foam adds to the compressive strength of the system, but more importantly, it adds to the vibration dampening for the blades. Figure 7(f) shows the finished blade as it was removed from the mould before surface finishing. Figure 7(g) is an image of the blades with the top coat on and connected to the hub. The blade hub is coupled to the low-speed shaft, which transfers the power from the wind through the gearbox to the generator.

6.3 Self-healing study in rotating blades

6.3.1 Testing blades

Wind turbine blades with and without self-healing devices was tested in the wind tunnel test by assessing high air speeds up to 20–30 m/s; stress is measured along a prototype wind turbine blade.

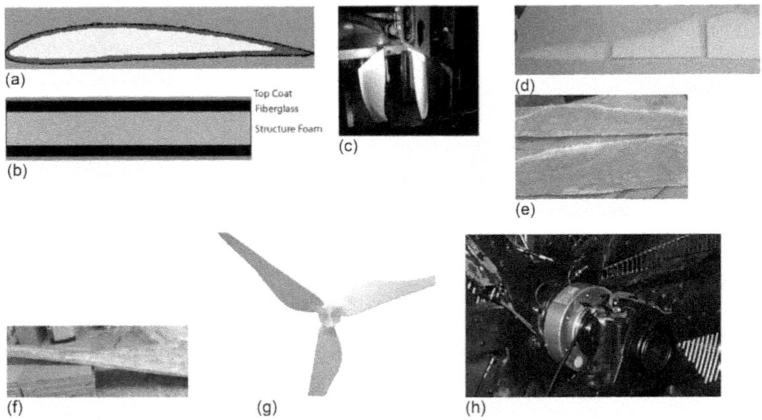

Figure 7: Blade formation process. (a) Blade cross-section, (b) blade composition, (c) production of the rapid prototyped blade, (d) rapid prototyped blade, (e) blade mould, (f) blade out of the mould, (g) finished blade and (h) slip wing installation at the blade shaft (UWM Wind Tunnel Lab).

One strategy to achieve self-healing functionality in polymer-based composites is to embed hollow vessels or capillaries containing a reactive healing liquid inside the material during processing. Then, when the material is damaged, cracks propagate through the material, rupturing the vessels or capillaries and releasing the reactive healing agent into the crack where it solidifies and repairs the blade structure. The healing process is triggered only by the crack propagation and requires no outside intervention [30].

One limitation to the microencapsulated healing agent systems and the phase-separated systems is that repeated healing is possible after a first healing event only if continued liquid is present at the damage region. It is not possible to know when the liquid healing agents have been entirely consumed. The use of hollow tubes or fibres, as shown in Fig. 8(b), is the approach that delivers larger amounts of a liquid healing agent to the crack plane. The hollow fibres are multifunctional, as the fibres store the liquid healing agent while they simultaneously provide structural reinforcement. Composite panels are subjected to low energy (~80 J) impact to rupture the hollow fibres and release the healing agent, followed by compression after impact testing. These systems, however, failed to adequately release the healing agents into the crack plane without the use of an external vacuum and heat. The curing of cyanoacrylate systems on contact with the mouth of the capillary and the high viscosities for the epoxy systems precluded significant self-healing from being realised [24]. Fortunately, on a rotating blade, a crack usually occurs at a very high local speed along the blade with approaching wind velocity over 20 m/s, where the pressure becomes significantly low due to the Venturi effect. Such a low-pressure helps to exude the healing agent sufficiently when the fibre tube ruptures.

We first need to generate the stress distribution for every model of the blade with and without self-healing embedded, as was done in the preliminary study, which shows preliminary stress distributions over wind turbine blades. A basic self-healing design and determination of the locations of self-healing fibre tube inside the blade can be made based on this stress distribution. One such arrangement is shown in Fig. 8. The self-healing agent within hollow fibres was arranged along with the strengthening fibres with a catalyst distributed around these fibres and was installed in a wind turbine blade (see Fig. 8(a)). A composite with hollow tubes containing a healing agent surrounded with catalysts is shown in Fig. 8(b) with CFD confirmation (Fig. 8(c)) and microstructure photo (Fig. 8(d)).

The self-healing embedded blade was investigated by increasing the air speed in the wind tunnel up to 30 m/s, while the stress levels through the blade are monitored. After the wind fan is stopped, the self-healing embedded blades are analysed for micro-structural deformation.

For the healing agent, Bond *et al.* [31] developed a process to optimise the production of hollow glass fibres and have since used those fibres (filled with liquid healing agents or dyes) in polymer matrix composite for damage detection and self-repair. This type of fibre tube is an ideal healing agent for wind turbine blade applications.

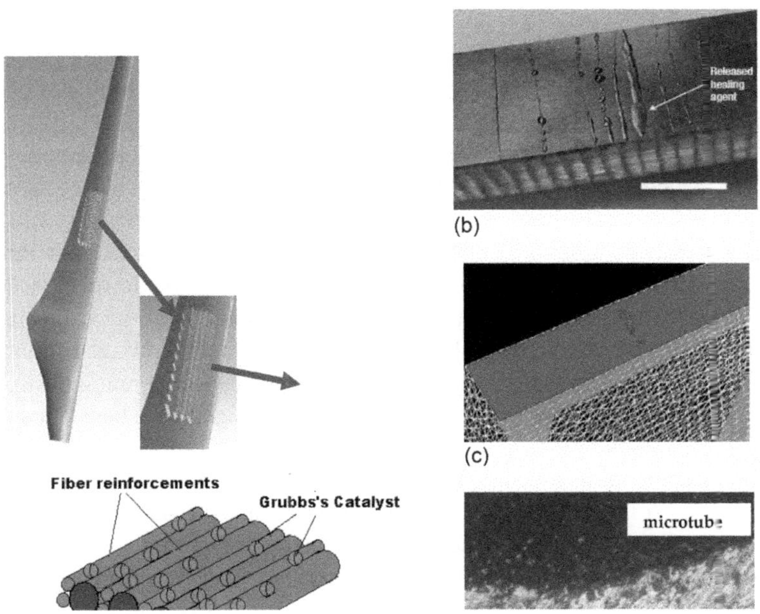

Figure 8. Test wind turbine blades. (a) Blade with self-healing devices, (b) hollow fibre with healing agent used by the PI's group (UWM Composite Lab) [33], (c) CFD results of (b) made by the PI's group (UWM Composite Lab) [34] and (d) optimum microstructure of the interface (UWM Composite Lab).

6.3.2 Catalyst

Two generations of the catalyst, benzylidene-bis (tricyclohexylphosphine) dichlororuthenium and benzylidene [1,3-bis (2,4,6-trimethylphenyl)-2-imidazolidinylidene] dichloro (tricyclohexylphosphine)] ruthenium (Sigma–Aldrich, USA) were used for this study. Both Grubb's first- and second-generation catalysts (see Fig. 4) are tolerant of many functional groups and compatible with a wide range of solvents, and are stable to air and moisture, which makes handling easy [32]. In self-healing applications, the suspended catalyst must remain active for rheological behaviour of ROMP with the healing agent after processing and curing of the thermosetting matrix [33].

6.4 Thermoplastic turbine blades

Usage of thermoplastics can be thought as an alternative and improved option when compared with thermoset plastics. Usage of thermoplastics gives an edge over thermoset plastics, which can be used to produce larger and thicker blades. A spar can be included to give the structural stiffness for the blade due to the increase in size. Thermoplastics can be processed by the conventional melt processing or by newer reactive processing. Reactive processing gives better properties when compared with melt processing. The disadvantages of melt processing are: (1) poor fatigue performance due to poor fibre-to-matrix bond; (2) requires introduction of new and expensive technology; (3) high material costs due to the need for intermediate processing steps; (4) high processing temperatures (>200°C); (5) melt processing limits achievable part size and thickness. For comparison, the advantages of reactive processing can bring better impact properties and do not turn brittle at low temperatures. Also, it has unlimited shelf-life of raw material and environmental & economic benefits from fully recyclable blades. These advantages over thermoset plastics motivated us to study on thermoplastic wind turbine blades.

Vacuum-assisted resin transfer moulding (VARTM) is the vacuum infusion technique currently used in the industry for wind turbine blade manufacturing. VARTM basically uses vacuum at the reinforcement material to suck the polymer or monomer and form a uniform layer of composite material. Usage of this technique to produce thermoplastics adds as an advantage as the current industry does not have to undergo complete retooling during the changeover. The advantages of using VARTM technique to manufacture blades are: (1) commonly used technology for blade manufacturing; (2) manufactured using reactive processing; (3) manufacturing of larger, thicker and more integrated thermoplastic composite parts; (4) improved chemical bonding due to in situ polymerisation of the matrix around the fibres and (5) manufacturing of parts directly from the monomer

6.4.1 SEM test

The cross-sections of each sample were investigated under SEM after tensile test. The results are shown in Fig. 11. The (a) shows the cross-section of sample 1 (90°/0°, 15 layers), we can notice a completely straight line along the break down section. When we decrease the layer number of glass fibre to six layers but keep

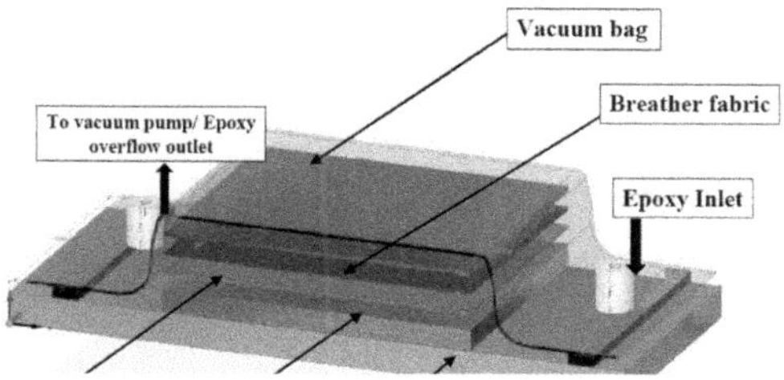

Figure 9: Standard mould design.

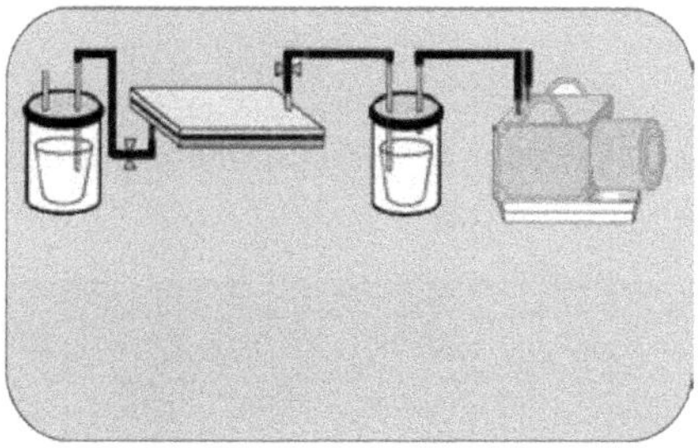

Figure 10: Schematic of vacuum-assisted resin transfer moulding (VARTM).

the same glass fibre orientation, the break down section begins to show the cracks on the edge. If we further change the configuration of glass fibre layer, this time we change the fibre orientation into 90°/45° but keep the same number of layers, under SEM more cracks on the edge was observed compared with other two samples. This may indicate the stronger elongation resistance in sample 3.

6.4.2 Tensile test

To determine which sample would be the best for real-world applications, it must have a high maximum stress and elongation. Sample 1 had the lowest maximum stress and lowest elongation, even though it had more layers than the other two layers. This was due to an error in the fabrication of the original sample. When doing the vacuum bagging process, the vacuum bag was not sealed properly, so excess resin was allowed to cure on the top of the glass fibre layers. This resulted in the sample having a lower volume fraction of fibre than the other samples, which resulted in the lowest stress bearing capability.

Sample 2 had the highest maximum stress due to its anisotropic properties. All of the layers were oriented parallel to the tensile test load, so most of the fibres in the sample carried the load. Sample 3 had slightly less maximum stress than sample 2. This was because sample 3 had quasi-isotropic properties, which allows the sample to have the same properties in any direction. The sample had 90°/45° alternating layers, so only half the layers were lined up to the optimal position to carry the load from the tensile test. Sample 3's alternating layers allowed it to have higher ductility because the 45° layers were able to slip between the 90° layers.

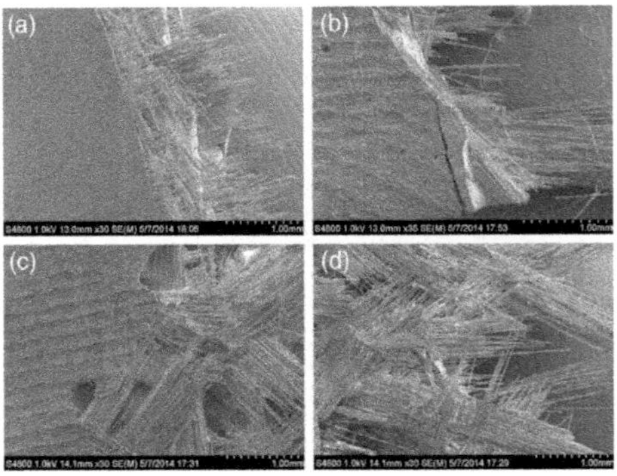

Figure 11: SEM image. (a) Sample1; (b) sample 2 and (c and d) sample 3.

Figure 12: A wind turbine spins amid flaxen wheat fields in Edgeley, North Dakota. Wind turbines convert kinetic energy from wind into electrical, thermal or mechanical energy with virtually no harm to the environment. Photograph by Sarah Leen [35].

Table 1: Tensile test results.

	Cross-Sectional Area (in^2)	Max Load (lb)	Maximum Stress (lb/in^2)	Original Guage Length (in)	Final Guage Lenth (in)	Elongation (%)
Sample 1	0.1204	1114	9256	2.131	2.172	1.92
Sample 2	0.0529	1099	20769	2.105	2.225	5.7
Sample 3	0.0522	960	18375	2.24	2.432	8.57

Even though sample 2 had the highest maximum stress capability for this test, it would not have as much if it were oriented 45°. This limitation in directional load makes uni-directional samples sub-par compared with the 90°/45° because the 90°/45° orientation allows for the same load to be carried in all directions (Table 1).

7 Conclusions

This chapter summarises a method to apply the self-healing technology to a wind turbine blade, which can be used for a future turbine blade failure prevention method. Wind turbines have no harm to the environment (Fig. 12), and it is one large candidate for energy production device for our future.

References

[1] American Wind Energy Association (AWEA) Report, January 2010.

[2] The White House, http://georgewbush-whitehouse.archives.gov/ask/20061011.html (accessed 15 February 2012).

[3] Wind Power America, US Department of Energy, March 2010, http://www.windpoweringamerica.gov/wind_maps.asp (accessed 15 February 2012).

[4] Amano, R.S., Avdeev, I., Malloy, R.J. & Shams, M. Z., Power, structural and noise performance tests on a different wind turbine rotor blade design. *International Journal of Sustainable Energy*, doi: 10.1080/14786451.2011.599491. Available online: 18 July 2011.

[5] Martin, J.C., Barroso, A., Paris, F. & Canas, J., Study of fatigue damage in wind turbine blades. *Engineering Failure Analysis*, **16(2)**, pp. 656–668, 2009.

[6] Amano, R.S., Avdeev, I., Mohan Das, P. & Zunaid Shams, M., Power performance tests on a different wind turbine rotor blade design. *Proceedings of ASME POWER2011 Conference, POWER2011-55226*, July 12–14, 2011, Denver, CO.

[7] Dry, C.M. & Sottos, N.R., Passive smart self-repair in polymer matrix composite materials. *North American Conference on Smart Structures and Materials*, Albuquerque, USA, pp. 438–444, 31 January to 4 February 1993.

[8] Gosh, S.K., *Self-Healing Materials*, Wiley-VCH, Weinheim.

[9] Martin, J.C., Barroso, A., Paris, F. & Canas, J., Study of fatigue damage in wind turbine blades. *Engineering Failure Analysis*, **16(2)**, pp. 656–668, 2009.

[10] Chen, X., Dam, M.A., Ono, K., Mal, A.K., Shen, H., Nutt, S.R., Sheran, K. & Wudl, F., A thermally re-mendable cross-linked polymeric material science. *Science*, **295(5560)**, pp. 1698–1702, 2002.

[11] Zako, M. & Takano, N., Intelligent material systems using epoxy particles to repair microcracks and delamination damage in GFRP. *Journal of International Materials Systems & Structures*, **10(10)**, pp. 836–841, 1999.

[12] Hayes, S.A., Jones, F.R., Marhiya, K. & Zhang, W. Self-healing composite materials. *Proceedings of 15th International Conference on Composite Materials*, Durban, South Africa, 27 June to 01 July 2005.

[13] White, S.R., Sottos, N.R., Geubelle, P.H., Moore, J.S., Kessler, M.R., Sriram, S.R., Brown, E.N. & Viswanathan, S., Autonomic healing of polymer composites. *Nature*, **409**, pp. 794–797, 2001.

[14] Van der Zwaag, S., An introduction to material design principles: damage prevention versus damage management. *Self Healing Materials*, ed. S. Van der Zwaag, Springer: the Netherlands, 2007.

[15] Blaiszik, B.J., Kramer, S.L.B., Olugebefola, S.C., Moore, J.S., Sottos, N.R. & White, S.R., Self-healing polymers and composites. *Annual Review of Materials Research*, **40**, 2010, pp. 179–211.

[16] Thies, C., Microencapsulation. In *Encyclopedia of Polymer Science and Engineering*, vol. 9. John Wiley and Sons, Inc.: New York, pp. 724–745, 1987.

[17] Jung, D., Hegeman, A., Sottos, N.R., Geubelle, P.H. & White, S.R., Self-healing composites using embedded microspheres. *American Society of Mechanical Engineers Materials Division*, **80**, pp. 265–275, 1997.

[18] Trask, R.S., Williams, H.R. & Bond, I.P., Self-healing polymer composites: mimicking nature to enhance performance. *Bioinspiration & Biomimetics*, **2**, pp. 1–9.

[19] Amano, R.S., Martinez Lucci, J., Rohatgi, P., Schultz, B. & Ruzek, A., Design and demonstration of self-healing behavior in a lead-free solder alloy. *7th AIAA International Energy Conversion Engineering Conference*, Denver, CO, 2–5 August 2009.

[20] Lucci, J.M., Amano, R.S. & Rohatgi, P.K., Computational analysis of self-healing in a polymer matrix with microvascular networks. *Proceedings of the ASME 2008 International Design Engineering Technical Conferences & Computers and Information in Engineering Conference*, DETC2008-50148, New York, NY, 3–6 August 2008.

[21] Toohey, K.S., White, S.R. & Sottos, N.R., Microvascular networks for self-healing polymer coatings. *Fifteenth United States National Congress of Theoretical and Applied Mechanics*, Boulder, CO, USA, 25–31 June 2006.

[22] Williams, H.R., Trask, R.S. & Bond, I.P., Vascular self-healing composite sandwich structures. *Fifteenth United States National Congress of Theoretical and Applied Mechanics*, Boulder, CO, USA, 25–31 June 2006.

[23] Lucci, J.M., Amano, R.S., Rohatgi, P.K. & Schultz, B., Self-healing in an aluminum alloy reinforced with microtubes. *Proceedings of the ASME 2008 Energy Nano Conference*, ENIC2008-53011, Jacksonville, FL, 10–14 August 2008.

[24] Amano, R.S., Martinez Lucci, J., Rohatgi, P. & Schultz, B., Design and demonstration of self-healing behavior in a lead-free solder alloy reinforced with carbon fiber microtubes. *2nd International Conference on Self-Healing Materials*, Chicago, IL, June 28 to July 1, 2009.

[25] Grubbs, R.H., Olefin metathesis catalysts for the preparation of molecules and materials, *Nobel Lecture*, 8 December 2005.

[26] Xu, C., Amano, R.S. & Perez, R.A., Experimental study on unsteady pressure field investigation of an axial fan: inlet and outlet unsteady pressure field measurement. *International Journal of Rotating Machinery*, **8(6)**, pp. 375–383, 2002.

[27] Xu, C. & Amano, R.S., Computational analysis of swept compressor rotor blade. *International Journal for Computational Methods in Engineering Science & Mechanics*, **9**, pp. 374–382, 2008.

[28] Xu, C. & Amano, R.S., Computational analysis of pitch-width effects on the secondary flows of turbine blades. *Computational Mechanics*, **34(2)**, pp. 111–120, 2004.

[29] Amano, R.S. & Sunden, B., *Thermal Engineering in Power Systems*. WIT Press, 416 pp, 2008.

[30] Amano, R.S., Martinez Lucci, J., Rohatgi, P., Schultz, B. & Ruzek, A., Design and demonstration of self-healing behavior in a polymer. *7th AIAA International Energy Conversion Engineering Conference*, Denver, CO, 2–5 August 2009.

[31] Hucker, M., Bond, I., Foreman, A. & Hudd, J., Optimisation of hollow glass fibres and their composites. *Advanced Composites Letters*, **8(4)**, pp. 181–189, 1999.

[32] Trnka, T.M. & Grubbs, R.H., The development of L2X2Ru=CHR olefin metathesis catalysts: an organometallic success story. *Accounts of Chemical Research*, 34(1), pp. 18–19, 2001.

[33] Martinez Lucci, J., Amano, R.S. & Rohatgi, P.K., Self-healing in an aluminum alloy reinforced with carbon fiber microtubes. *AIAA Aerospace Sciences Meeting*, AIAA-2009-1019, Orlando, FL, 5–8 January 2009.

[34] Guntur, K.S., Amano, R.S., Martinez Lucci, J, Rohatgi, P. & Schultz, B., Self-healing technology for compressor and turbine blades. *ASME Turbo & Expo*, GT2009-59103, Orlando, FL, 8–12 June 2009.

[35] National Geographic, http://science.nationalgeographic.com/science/space/universe/after-oil-energy/

[36] Human body net vascular system Paradoja7.com, http://www.paradoja7.com/pictures-of-human-body/

www.ingramcontent.com/pod-product-compliance
Ingram Content Group UK Ltd.
Pitfield, Milton Keynes, MK11 3LW, UK
UKHW020819290926
15470UKWH00003B/1251

9 781784 660048